THE ILLUSTRATED ENCYCLOPEDIA OF

BIRDS

THE ILLUSTRATED ENCYCLOPEDIA OF BIRDS

By Dr Jan Hanzák
and Dr Jiří Formánek

Edited by
Laura Steward

PRC

PICTURE ACKNOWLEDGEMENTS

(Figures refer to numbers of illustrations)

Balát F.: 25, 31, 76, 86, 180, 210, 217, 218, 232, 235, 323, 378, 381, 395 **Beretzk P.:** 227 **Boisson J. Y.** (Jacana): 7, 208 **Brosset** (Jacana): 28, 39 **Bruce Coleman Ltd.**: 104, 110, 167, 298, 329, 350, 362, 412, 446, 452 **Burton J.** (Coleman): 444 **Čeřovský J.:** 143, 364 **Christiansen A.:** 99, 197 **Čihař J.:** 139, 172 **Claye M.** (Jacana): 37 **Čtyroký P.:** 189, 190, 203, 307, 311, 349, 365, 372, 469, 470 **Dawson L. R.** (Coleman): 163, 297 **Dubois J. L. S.** (Jacana): 249, 250, 317 **Eliott** (Jacana): 49 **Fermie** (Jacana): 27 **Firsova L. V.:** 212, 215, 229, 234, 236, 237 **Formánek J.:** 11, 12, 13, 44, 45, 64, 68, 74, 78, 81, 91, 92, 95, 96, 100, 101, 102, 129, 137, 142, 157, 158, 170, 171, 185, 186, 193, 194, 196, 200, 201, 202, 223, 270, 271, 272, 274, 275, 276, 280, 281, 282, 287, 288, 289, 292, 293, 324, 325, 337, 347, 348, 366, 368, 369, 373, 374, 388, 394, 399, 425, 429, 437, 474, 482 **Fréderic** (Jacana): 240 **Fritz B.** (Jacana): 20 **Hanzák J.:** 5, 22, 23, 29, 33, 47, 48, 53, 56, 62, 63, 70, 72, 73, 82, 83, 84, 134, 144, 156, 178, 184, 187, 188, 192, 195, 204, 213, 216, 220, 221, 230, 231, 233, 340, 376, 403, 406, 421, 431, 442, 466 **Harris M. P.** (Coleman): 445 **Holeček Z.:** 90, 145, 151, 214, 247, 456, 471 **Hosking E.:** 286, 413 **Humpál Z.:** 89 **Jackson P.** (Coleman): 385 **Judin K. A.:** 199, 222, 402, 457, 472 **Klápště J.:** 19, 42, 67, 88, 183, 224, 252, 319, 357, 358 **Korba P.:** 41 **Krečmar A. V.:** 9, 191 **Lane Frank W.:** 59, 133, 246, 290, 294, 301, 328, 355, 410, 416, 427, 447, 453, 463, 464 **Laubscher C.** (Coleman): 386 **Lindberg P.:** 8 **Moll K. H.:** 10, 66, 105, 106, 131, 164, 165, 198, 360, 367, 454, 462 **Myers N.** (Coleman): 226 **Nejfeldt I. A.:** 273, 296, 300, 384, 390, 391, 392, 408, 422, 459 **Pavlík P.:** 40, 43, 57, 58, 65, 69, 98, 114, 116, 118, 119, 120, 128, 130, 135, 136, 168, 242, 245, 277, 291, 295, 299, 305, 306, 308, 309, 314, 326, 327, 352, 353, 354, 356, 359, 382, 383, 387, 389, 393, 396, 397, 398, 400, 401, 404, 405, 407, 409, 411, 415, 417, 418, 419, 420, 423, 424, 426, 428, 430, 432, 434, 455, 458, 460, 465 **Podpěra P.:** 478 **Portěnko L. A.:** 211 **Prevost** (Jacana): 6 **Raux F.** (Jacana): 24 **Rivarola H.** (Coleman): 279, 304 **Rys J.:** 60, 94, 140, 148, 205, 207 **Šálek J.:** 117, 125, 138, 146, 150, 162, 243, 255, 256, 257, 262, 263, 267, 268, 302, 363, 375, 380, 433, 450, 467, 468, 479, 480, 481 **Schraml W.** (Jacana): 342 **Schröder H.:** 126, 361 **Seget J.:** 3, 4, 14, 15, 16, 17, 18, 21, 30, 32, 34, 35, 46, 51, 52, 54, 55, 61, 71, 79, 80, 85, 87, 93, 108, 109, 111, 112, 115, 121, 122, 124, 127, 132, 147, 149, 152, 153, 154, 155, 159, 160, 161, 166, 169, 175, 176, 179, 181, 206, 228, 238, 239, 241, 244, 248, 251, 253, 254, 258, 259, 260, 261, 264, 265, 266, 269, 283, 284, 285, 310, 312, 313, 316, 318, 321, 322, 330, 331, 332, 334, 335, 336, 338, 339, 377, 414, 435, 436, 439, 440, 441, 443, 448, 449, 451, 461, 475, 477 **Šibněv J.:** 278, 379 **Simon J.** (Coleman): 303 **Stalla F.:** 38, 209 **Staněk V. J.:** 36, 50, 113, 123, 177, 320, 333, 351, 476 **Stivens D.:** 173 **Suchomel:** 75, 77, 438 **Suinot** (Jacana): 1 **Summ P.** (Jacana): 182 **Tollu B.** (Jacana): 2, 26 **Tomalin N.** (Coleman): 315 **Tomkovič P.:** 219 **Varin J. P.** (Jacana): 473 **Veselovský Z.:** 97, 103, 107, 343, 344, 370, 371 **Visage A.** (Jacana): 341, 345, 346 **Vlasák P.:** 141 **Wolf J.:** 174 **van Wormer J.** (Coleman): 225
Photographs on pp. 6 and 15 are by J. Formánek, on pp. 13 and 19 by H. Schröder.

Text by Dr Jan Hanzák and Dr Jiří Formánek
Translated by Olga Kuthanová
Graphic design by Karel Drchal

This edition published 1992, reprinted 1993
by The Promotional Reprint Co Ltd,
Deacon House, 65 Old Church Street,
London SW3 5BS, produced exclusively for
Angus & Robertson in Australia and
Smithbooks in Canada.

ISBN 1 85648 070 4
Printed in Slovakia by Neografia, Martin
3/11/02/51-05

CONTENTS

THE COLOURFUL WORLD OF BIRDS

Man has perhaps a greater interest in birds than in any other group of animals. These quick and agile creatures which are masters of the air have a rich instinctive life, interesting behaviour and exceptional vocal abilities. Without exaggeration they are an ornament of nature's realm. They are to be found in all geographic latitudes. A vast number of species brighten the tropical and subtropical zones; they are also plentiful in the temperate zones and their lovely appearance and ringing notes brighten even the most inhospitable reaches of the high mountains and circumpolar regions. It is difficult to imagine our woodlands, fields and meadows, lakes and rivers and even steppes and deserts without these feathered inhabitants, and many are the characteristics of these creatures that endear them to the heart of man.

Today's hectic pace and overmechanized age spur man to seek relief after a long day's work in observing nature's creatures in their natural environment, even though – unfortunately – it has often suffered from the inroads of civilization. That is why man's interest in birds and their way of life has grown so markedly. In all sorts of unusual places one can come across people with fieldglasses, or simply their own keen eyes, notebook in hand, watching the birds about them, listening to their song and trying to penetrate the secrets of their way of life. As the interest in bird-watching grows so also does the number of books and other publications designed to improve people's knowledge about birds or at least to make them more familiar with the world of birds through pictures and photographs.

This book gives a brief survey of the birds of the whole world, their varied types and environments. Unlike most studies of this type it purposely avoids a systematic classification of the individual species designed to present their family and evolutionary relationships in accordance with one of the presently acknowledged systems. Instead it presents to the reader examples of interesting species arranged in separate, thematically diversified but complete chapters that are not always in accord with the natural system. The book's main purpose is to acquaint the reader with the beauty of shape and colour as well as the wealth of forms to be found in the world of birds populating the earth and to call attention to certain interesting aspects of their life and habits.

ABOUT BIRDS IN GENERAL

Birds evolved from reptiles millions of years ago and to this day they have a number of features in common. As in reptiles, glands are almost completely absent from the skin, the skull is attached to the spinal column by a single joint, and the arrangement of the excretory and reproductive organs as well as the growth of the embryo show a marked resemblance to the reptilian pattern. Birds differ from reptiles, however, in having a constant body temperature (an average of 41°C) which is not dependent on the temperature of the environment. They are superbly adapted to life in the air. The few flightless species doubtless evolved from birds capable of flight for they have retained a number of anatomical characteristics of their flying kin. A bird's body is covered with feathers and the forelimbs have been transformed into wings, with only the legs serving for movement on firm ground. The feet have four or three toes, exceptionally only two (the Ostrich). In place of jaws the bird has a horny bill, which, apart from rare exceptions among mammals, is likewise a unique feature of birds. The heart is the same as in mammals. It is divided into two parts (left and right), each with two chambers, the one side pumping blood into the arteries and the other side collecting it from the veins. The lungs are fairly small and limited in their ability to expand and contract. The bronchial tubes pass through the lungs and terminate in large, thin-walled air-sacs. Basically there are three pairs which branch out into various parts of the body including the bones. They serve as reservoirs of air for breathing, especially in flight. Like reptiles,

BIRD TOPOGRAPHY

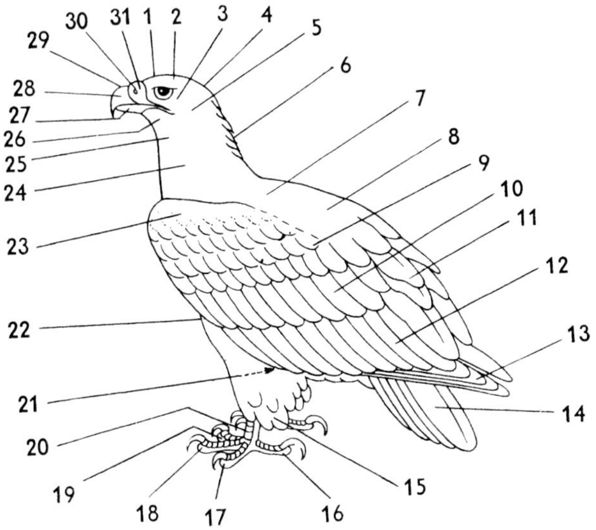

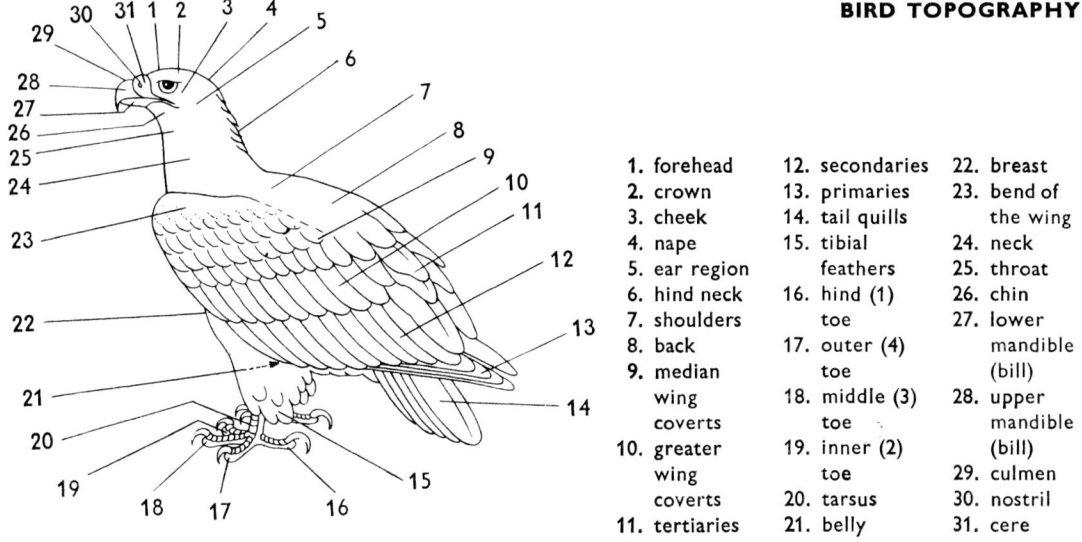

1. forehead
2. crown
3. cheek
4. nape
5. ear region
6. hind neck
7. shoulders
8. back
9. median wing coverts
10. greater wing coverts
11. tertiaries
12. secondaries
13. primaries
14. tail quills
15. tibial feathers
16. hind (1) toe
17. outer (4) toe
18. middle (3) toe
19. inner (2) toe
20. tarsus
21. belly
22. breast
23. bend of the wing
24. neck
25. throat
26. chin
27. lower mandible (bill)
28. upper mandible (bill)
29. culmen
30. nostril
31. cere

birds reproduce by laying eggs, which, however, have a firm calcareous shell. The embryo usually develops under the influence of the body heat of the adult birds, which incubate the eggs for a certain period and when the young hatch continue to care for them for some time.

BODY COVERING

Birds are the only creatures to have a coat of feathers, which are believed to have evolved over a long period of complicated transformation from the scales of the reptile. Feathers are composed of a horny substance which is the product of the skin. Their structure is delicate and light. Feathers serve partly as protection, mainly as insulators, and some serve to control flight (wing quills, tail quills). The shape of the body is determined by the contour feathers which possess the characteristic quill, rachis and web. The wing and tail quills, which have a very strong web, developed from the ordinary contour feathers. Another special type is the down feathers which have short, soft quills and, instead of a connected web, soft rays and barbs. They form a warm, downy layer underneath the contour feathers of adult birds and the first coat of most young birds on hatching. In some birds that have a reduced uropygial gland (e.g. parrots, cranes, herons) the down feathers continually disintegrate at the tip into powder that to a certain degree serves as a substitute for the secretion of the uropygial gland with which other birds' feathers are oiled. Only in the screamers, the penguins and the cursorial birds do the feathers grow continuously over the whole body surface. In all other birds they grow in certain definite tracts called pterylae. The intervening spaces, called apteria, however, are covered by the flight feathers growing at the edges of the pterylae. The bird's coat of feathers is not as thick as the fur of a mammal. Small species of hummingbirds have about a thousand feathers on their tiny bodies, swans some twenty to thirty thousand. The colour of a bird's plumage is produced by the structure of the feather and by pigments. Most important are the melanins, which produce black to pale brown hues, and secondly carotenoids, which produce red and yellow tones, various other hues resulting from the combinations of these pigments. Birds do not have any blue or purple pigments. Where these colours occur they are produced by the structure of the fine horny layers of the feather and the refraction of light.

Plumage is replaced regularly by a process called moulting. Most birds moult twice a year. In summer or autumn, following nesting, the bird changes its entire plumage. Another partial moult occurs in spring when the birds acquire the more colourful, so-called nuptial plumage. Some birds, however, moult once in two years while others moult three times yearly. During the moult the wing and tail quills are shed successively so that the bird does not lose the power of flight. Only geese, ducks, cranes and some other birds shed them all at once so that they go through a brief flightless period when they must remain hidden in the vegetation.

FEEDING

Another horny part of the bird's body is the bill, or rather its outer covering. The bill varies greatly in shape and indicates the method whereby the given bird feeds. It does not serve to chew food like the teeth of mammals. Birds that feed on seeds have a simple, cone-shaped bill adapted for gathering or splitting seeds. Insectivorous birds have a finer, awl-shaped bill, the hooked bill of the parrot serves to crack hard fruits and as a support when climbing, the sharp hooked beak of owls and raptors for tearing the flesh of their prey, the flat bill of most waterfowl for sifting food from the water in which they wade. The long pointed beak is a characteristic feature of birds that catch fish. Proportionately longest are the bills of birds that forage for food in mud and swamps. There are also down-curved and even up-curved beaks. The diversity of shape makes it possible to utilize various food sources and hunt prey even under difficult conditions — underwater, in crevices and even to dig it out of hard wood. The food, either whole or broken up by the bill, passes into the gullet. The narrow tongue, covered with a horny membrane, has apart from a few exceptions only a secondary function in the process. Before passing on to the stomach the food is usually retained for a time either in the throat pouch or crop, which is actually an enlarged section of the gullet. Here the food is not digested but merely softened. The crop is especially developed in seed-eating birds; in some groups it is only imperfectly developed or absent altogether. Protein digestion takes place in the antechamber of the stomach. The other part of the stomach, the gizzard, with thick muscular walls, is where food is ground and digested, this action often being assisted by the stones or grit, instinctively swallowed by the birds. The intestine is usually shorter than in mammals, this being in line with the structural economy of the bird's body. Undigested matter such as the chitinous remains of insects, bones, hairs and feathers are regurgitated by the birds in the form of pellets. Birds consume great quantities of food for they burn up a great amount of energy. It is a general rule that small birds require proportionately more food than large birds. The daily amount consumed by hummingbirds, kinglets and wrens equals more than their body weight. The process of digestion in birds is very rapid, more

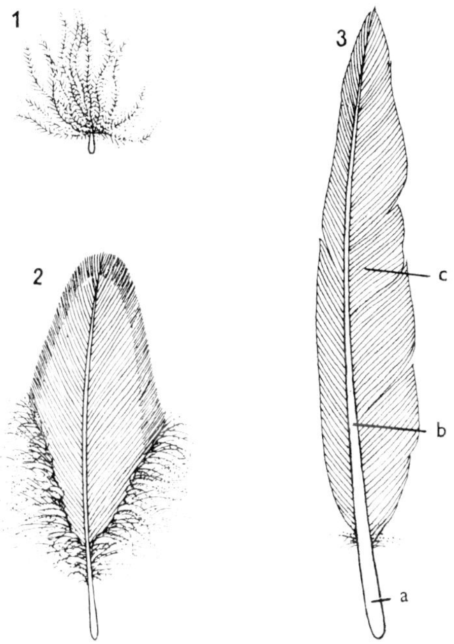

TYPES OF BIRD FEATHERS

1. down feather 2. contour feather 3. flight feather
a) quill b) rachis c) vane, web

TYPES OF BILLS

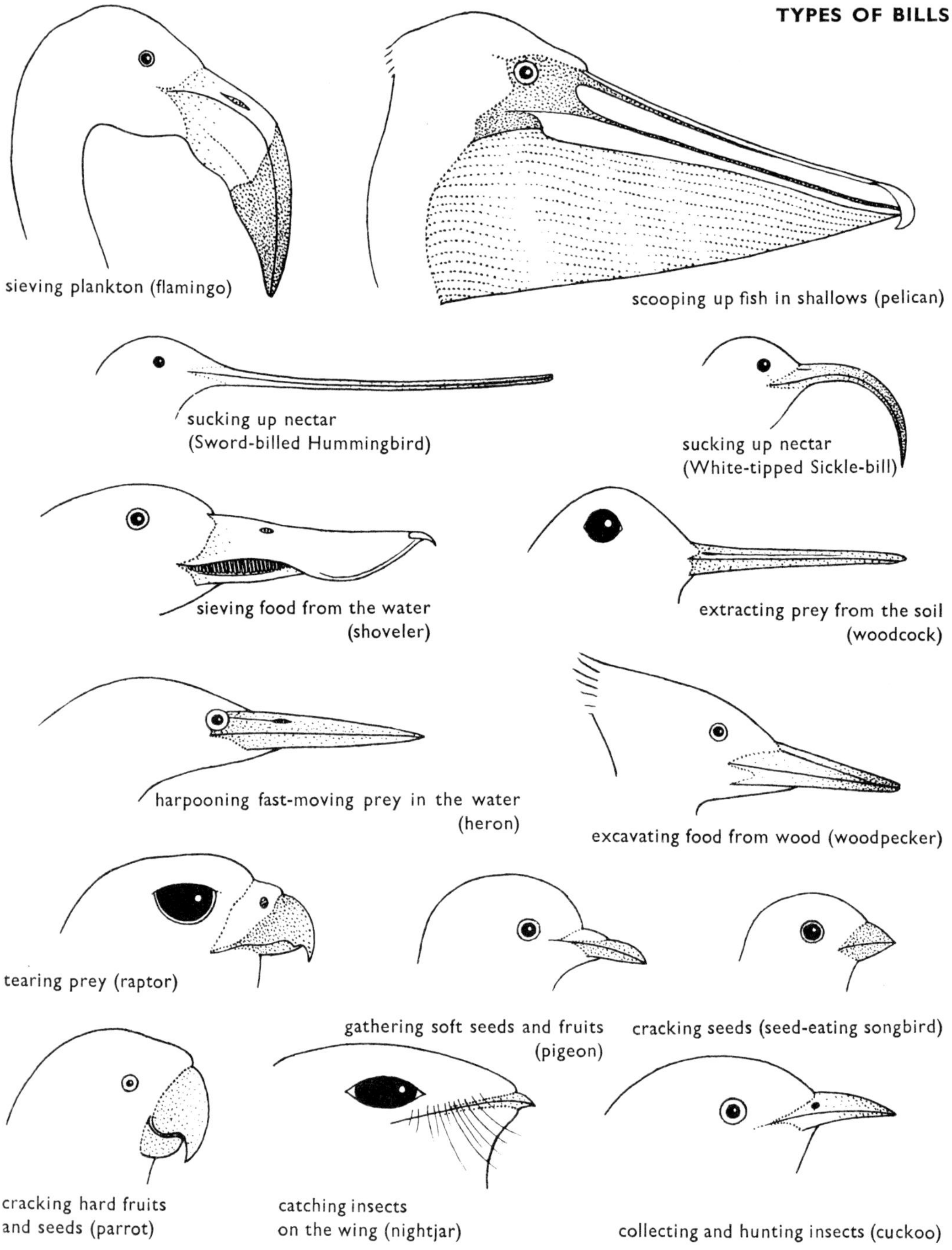

so than in any other vertebrates. In seeking their food and tracking their prey birds are mainly aided by their keen eyesight or hearing. Only few birds have a well-developed sense of smell and taste is the least developed of the senses.

LOCOMOTION

The evolution of birds as a group was marked mainly by the development of characteristics enabling them to master the air. The ability to fly depends first and foremost on lowering weight to the minimum and on the existence of a perfect supporting structure and power mechanism. The lowering of the bird's specific gravity has been achieved in several ways – instead of jaws with teeth there is a light toothless bill, there is no urinary bladder, the long bones are hollow and filled with air, the legs add little weight, and the feathers are light and efficient in structure. The most important instrument of flight are the feathered wings. In themselves, of course, they could not fulfil their function if they were not propelled by strong muscles attached to the robust keel of the breastbone. To fly and, at the same time, maintain body weight takes a great amount of energy, and this is instantly supplied by rapid digestion. The great exertion of flight is the reason why birds have a proportionately larger heart than mammals and a more rapid pulse-rate.

The method of flight shows marked diversity and depends first and foremost on the shape of the wings and the size of their surface area. The wing profile, however, is the same in all birds: convex above and concave below. In flight, air streaming over the body from front to back creates excess pressure underneath the wing which is about three times greater than the reduced pressure above the wing. The original form of flight from the evolutionary standpoint is simple, passive, gliding flight in which the bird does not flap its wings but is supported by the resistance of the air and descends slowly to the ground. Birds often alternate this with active flight. A gliding pigeon while slowly descending through a vertical height of ten metres actually travels a distance of ninety metres; better fliers, such as the eagle, travel a distance of 170 metres, the albatross even 200 metres, with the same ten-metre loss in height. When a bird flies head on into the wind or takes advantage of rising air currents the gliding flight may change to soaring, enabling the bird to acquire height without flapping its wings. Some birds of prey and albatrosses and storks are true masters of this art, whereas very few small birds use this form of flight since their wings have a very small surface area. The chief form of flight is flapping the wings, with the tips describing an elipse. The wings, powered by the breast muscles, propel the aerodynamically built body forward like oars propelling a boat on the water's surface. Another form of flight is hovering on the spot and, though less widespread, is often seen when the Kestrel, tern or Osprey is hunting for prey. A more elaborate form of the same type of flight is the hovering motion of the humming-bird, enabling this small bird to remain in one spot and extract the nectar or insects from flowers with its long bill without having to alight. The tips of the wings describe a figure of eight and the wing function in this instance may be compared with the propeller of a helicopter.

For most birds flight is essential to life. It enables them to make a rapid escape when danger threatens, to search for food, to settle new territories, and to migrate long distances. Many birds perform their courtship antics in the air. And without a doubt birds are the best travellers of all animals. Birds' flight performances were for a long time overestimated but they have since been revised. Recorded observations, however, continue to fill us with awe and admiration for these winged creatures. Remarkable, for instance, is the performance of the swift, which spends a full twelve to fourteen hours in the air flying at speeds of 65 to 145 kilometres per hour. Only recently it was proved that in attacking its prey the Peregrine

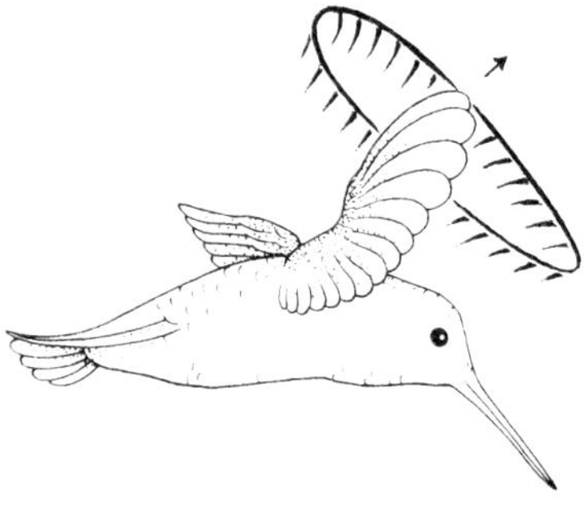

Hovering flight
of the hummingbird

Falcon attains short-term speeds of up to 280 kilometres per hour. This is an example of several peak performances, but even the average cruising speed of other birds is not negligible, being between 40 and 120 kilometres an hour. Most birds fly at a height of three hundred metres above the ground. Only during migration do larger birds exceed this limit. Instances of birds observed at heights above two thousand metres (geese even above eight thousand metres) are truly exceptional.

All birds dependent on water for their food are able to swim. Some do so only when resting on water and apart from webs on their toes have no other special adaptations for life on water. Quite different, however, is the case of birds that spend most of their lives in water or even dive: their feet have three or even all four toes joined by webs, though some water birds have only leathery lobes on either side of the toes. Good divers have bones that are not very pneumatic and are able to expel the air from the feathers by pressing them close to the body. Their specific gravity is greater than that of other birds and thus it is easier for them to overcome the resistance of the water when diving. When moving on the surface they usually propel themselves only with their feet, penguins being the one exception – for they propel themselves only by means of their wings. Underwater, birds propel themselves either with their feet or their wings; only penguins, alcids and certain tubenoses use the latter method, whereas other birds use their feet, the wings serving at best only for steering.

Birds that are very adept at swimming and diving are usually less expert on land. Divers, grebes and most alcids find travel on land quite difficult, the reason being that their feet are placed well at the back of the body and the tibial part of the leg has only limited movement. Similar limitations on land may be found among some of the top fliers. Otherwise all birds are capable of locomotion on land. The best runners are birds with no keel on the breastbone – Ostrich, Emu, Rhea, Cassowary, which have strong legs with few toes. With the exception of the Cassowary all are inhabitants of open expanses and it is no accident that even the good running birds of other groups, e.g.

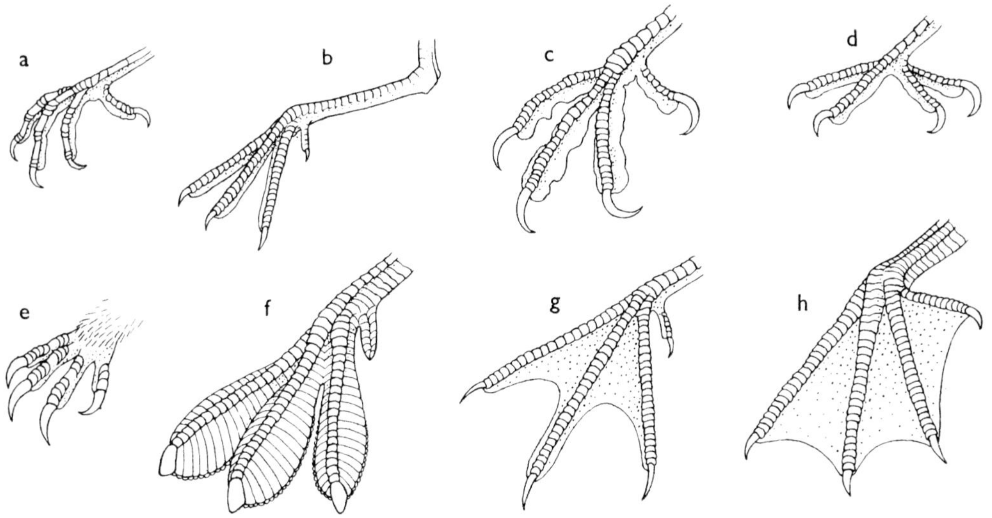

TYPES OF FEET

a) perching foot of arboreal birds
b) wading foot
c) raptor's foot
d) foot of clinging birds
e) gripping foot of the swift
f) grebe's foot
g) webbed foot of the tern
h) webbed foot of the cormorant

bustards and certain gallinaceous birds, are inhabitants of the steppes, semi-deserts and deserts. The ability to walk well and run quickly does not necessarily go hand in hand with long legs. Long-legged birds such as herons, storks or cranes, step along nicely but are unable to run fast. On the other hand it is interesting to note that among waders the plover is a better runner than the godwit or sandpiper, though the plover's legs are much shorter. Besides running, birds also travel by hopping. Why some birds walk and others hop is difficult to explain. Life in branches of trees has brought about numerous adaptations of the feet, e.g. sharp ridges on the underside of the toes, serrated combs on the claws, and mechanisms whereby the bird automatically maintains its foothold on the branch. In enumerating the various methods of locomotion one must not overlook climbing, which is a very common method among birds that live in trees. They have short feet with sharp claws, two toes pointing forward and two backward. When climbing tree trunks or cliffs most of these birds use their stiff tail feathers as a prop. This, of course, applies to those that climb head upward; nuthatches, however, often climb head downward.

Just as the beak reflects the type of food a bird eats and its method of procuring it, so the foot reflects its way of life and the shape is characteristic for each separate systematic group of birds.

REPRODUCTION

All birds produce offspring by laying eggs. Unlike the mammals the embryo does not develop inside the body of the female but in the egg after it is laid. As everyone knows, the egg chiefly comprises the yolk and the albumen, or egg-white.

The vital part of the egg is the yolk and the blastodisc, which is the actual embryo-forming cell and is the largest to be found among the vertebrates. The yolk and egg-white are enclosed in a colourless, translucent, papery membrane; at the rounded end of the egg this does not adhere

CROSS-SECTION OF NON-INCUBATED BIRD'S EGG

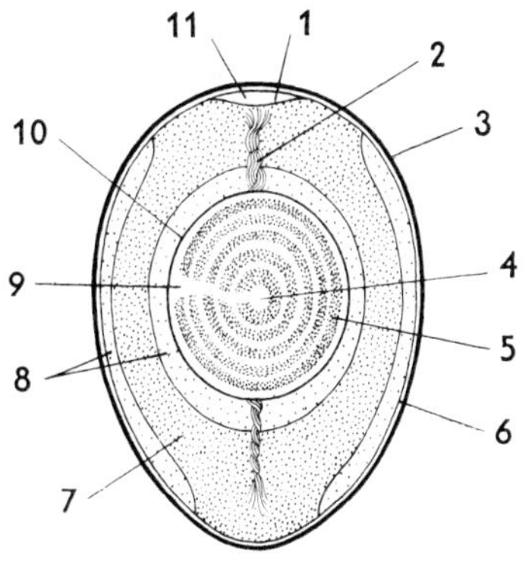

1. egg membrane
2. chalaza
3. shell
4. latebra
5. yellow yolk
6. shell membrane
7. gelatinous semi-solid albumen
8. viscous liquid albumen
9. blastodisc
10. vitelline membrane
11. air space

to the shell and an air space is formed that increases in size as the egg is incubated. The whole egg is enclosed in a porous shell which is composed chiefly of calcium carbonate with slight admixtures of salt and certain organic substances. Eggs are either white, a single colour or with various coloured markings. The largest egg of all extant birds is laid by the largest member of the avian realm – the Ostrich. It weighs about 1,600 grams and could hold twenty hen's eggs. The smallest eggs are laid by the hummingbirds, some of which are the smallest birds in existence; the egg weighs 0.25 grams. It is interesting to note that in proportion to their body weight small species of birds lay larger eggs than large species. The number of eggs is generally constant, not only for the given species but often for all the species of a given systematic group. The guillemots, together with some auklets, large gulls and birds of prey lay a single egg. Two eggs are the rule for pigeons, hummingbirds, cranes and divers. Waders lay three to four eggs. The largest number of eggs is laid by the gallinaceous birds, members of the family Anatidae and some songbirds. Clutches of more than twenty eggs are rare (Ostrich, partridge). Some species, despite their high rate of mortality, lay few eggs but supplement their numbers by having several broods a year.

Apart from a few exceptions the eggs are incubated by the parents, or at least one of the adult birds. Usually both take turns incubating at regular intervals (every few hours as a rule, but sometimes even every quarter-hour). In many cases the eggs are incubated by the female alone, but incubation exclusively by the male is extremely rare. The eggs need not be kept covered continuously and the adult birds sometimes leave them unguarded for quite long periods if they have to forage for food or preen their feathers. To develop successfully the embryo requires heat, which means that the body of the adult bird must be in direct contact with the egg during incubation. In order to facilitate the transfer of heat most nesting birds have featherless patches on the belly called brood-patches; the feathers either fall out by themselves or are plucked out by the birds. These patches are either entirely bare or covered with only a thin down. Ducks, however, do not have these bare patches. During the nesting period an exceptionally fine down grows in on the belly and breast of the female which she plucks to line the nest, thus preventing heat from escaping from the nest cup. Brood-patches appear either in the male or female, depending on which of the two incubates, and often in both.

There is no rule that the period of incubation depends on the size of the bird. For instance the Black Woodpecker and the Chaffinch incubate for the same length of time – twelve to fourteen days. Small songbirds require twelve to eighteen days for their young to hatch, gallinaceous birds twenty-one and eagles thirty-four to thirty-eight days. Fairly long is the incubation period of auklets, gannets, shearwaters and petrels, and albatrosses, some of which hold the record in the bird world – a full eighty days.

Not all birds incubate their eggs, but these are few. For instance, the megapodes inhabiting the damp forests of Australia and New Guinea incubate their eggs in what might be described as artificial hatcheries. The males rake together huge heaps of vegetation, soil and sand, piling up a mound in which the females lay their eggs. When it rains the vegetation starts to rot and to generate considerable heat which is sufficient to ensure successful incubation. To the question of what time of the year birds nest there is no unequivocal reply, for this depends primarily on the geographical location of the nesting grounds. In the temperate zone there are marked differences between the yearly seasons, some of which, with their unfavourable temperatures and lack of food, exclude the possibility of rearing young. In the tropics conditions are different and more propitious. In certain areas each species has a specific and limited nesting season characteristic of the given species and providing the optimum requirements for the development and rearing of the young. Birds that have a lengthy period of incubation and care for the young a long time after fledging begin nesting earlier than those with shorter periods of incubation and aftercare.

CARE OF THE YOUNG

No one stands by to help the young nestling enter the world. Piercing the shell is no easy task. This usually takes place at the rounded end of the egg, the young bird pecking its way out with the aid of the so-called 'egg-tooth'. This is a horny projection of the upper mandible that disappears shortly after hatching.

Birds differ in the degree of their development on hatching. Generally they are divided into two groups – nidifugous and nidicolous species. Birds of the first group generally nest on the ground or on water and include the running birds, gallinaceous birds, bustards, rails, ducks and geese. Their young leave the nest within a few hours to two to three days after hatching and are fairly independent. They are covered with a coat of fine down and are able to run about, in the case of water birds also to swim, from the very first day. They usually feed themselves, or else are guided to food by the parents so that they soon know what to eat and what not. For quite some time they continue to be watched over and guided by the parent birds to locations where food is to be found, warned against dangers and protected from enemies, and sheltered and kept warm under their wings at night and in cold weather.

The opposite of this group are the nidicolous birds which are the most numerous. They include

Two-day-old nidicolous nestling (songbird) and nidifugous chick (fowl)

the songbirds and most other orders. Their young are quite helpless on hatching, being blind and naked or covered only with a light down. They are entirely dependent on the parents, which feed them in the nest and often even for a long time after they leave it. The young require great quantities of food and so the adult birds are kept busy. The natural outcome of such conscientious feeding is the rapid growth of the young. Songbirds catch up with nidifugous birds of the same size within fourteen days and outweigh them soon after. The intensity of feeding the young of nidicolous species is truly remarkable. Tits and nuthatches bring their nestlings food 350 to 550 times a day and towards the end of the nesting period even more frequently. Some birds, e.g. gulls and terns, stand somewhere in between these two basic groups in the degree of development of the young in the first few days after hatching.

The methods of feeding the young are many. Some birds merely pass the food to the nestlings, others regurgitate it from the crop onto the nest or directly into the beaks of the young, some deliver the food whole, others partly digested or portioned. Pigeons feed their young so-called pigeon's milk – a cheesy curd regurgitated from the crop. Nidifugous birds, whose young start foraging for food themselves shortly after they hatch, are the ones with the least problems. Doubtless everyone has noted how busily chicks only a few days old scamper about in search of food.

When the young leave the nest this does not mean the end of the family relationship, which continues for some time after. Some birds of prey, geese, swans, cranes and other birds remain together in the family unit until the spring of the following year. On the other hand, small birds part from their young soon after fledging, the adult birds sometimes preparing to raise another brood. Here the family relationship is of short duration.

NESTING PARASITES

A special group of birds are those that do not raise their offspring themselves but entrust their eggs and the rearing of the young to other birds. This so-called nesting parasitism is found among certain weavers, bee-eaters, sunbirds, cuckoos and even ducks. This phenomenon is most highly developed in the cuckoo and is known in detail. The female cuckoo generally lays twelve to twenty eggs in one season. These are coloured the same as the eggs of the foster parents and are also the same size (they are extremely small in proportion to the size of the cuckoo). Their coloration shows a greater diversity than almost any other bird's, corresponding predominantly to the coloration of the eggs of the hosts which are insect-feeding birds. This means that over generations there have evolved certain biological races with eggs resembling those of the host species. Cuckoo eggs have an exceptionally thick shell to prevent them from breaking when they are dropped or hurriedly laid in the stranger's nest. A further adaptation is the shorter period of embryonic development. The young cuckoo hatches after only twelve and a half days and as the egg is almost always deposited in the nest when the host starts laying its own it hatches sooner or at the same time as the nestlings of the foster parent. The skin of the young cuckoo's back is extremely sensitive to contact with foreign objects and any such object in the nest – be it an egg or young nestling – is tumbled by the young cuckoo over the edge of the nest. Thus it remains the sole occupant and receives the full attention of its foster parents. It consumes as much food as five or six young songbirds and within three days doubles its weight. The foster parents bring food for a full five weeks to this strange nestling, which in the end is much larger than themselves.

BUILDING THE NEST

Building a nest is tied up with the development of the reproductive glands and is typically instinctive behaviour. The nest provides a temporary shelter not only for the eggs but also protection against danger and bad weather for the incubating birds as well as for the young nestlings. Only a few bird species do not build nests but lay their eggs either directly on the ground or in cavities. Nesting on the ground is the earliest form from the viewpoint of evolution and is very widespread. Water birds generally build their nests near water, some directly above the water's surface in a mass of vegetation; extreme examples are nests floating directly on the water (e.g. grebes). As a rule, however, nests are built in trees, bushes and taller plants. Some species of birds are satisfied with only a few twigs laid haphazardly in the forks of branches (certain pigeons), whereas others – and that includes the majority – build nice sturdy nests where branches join the trunk, or in the canopy on thick branches or even on their thinnest tips. Master architects include the Penduline Tit and many species of weaver. They build globular nests covered on all sides, and weavers even build communal nests.

A far safer method, however, is nesting in cavities, which is also very common among birds. Some species are content with merely the barest hint of a cavity and build their nests, for instance, in the hollow of a rotted tree stump or groove left by a broken branch, under a piece of peeling bark, in a shallow depression in sand or mud banks or in a pile of stones, so that the nest is not enclosed on all sides. Other species, on the other hand, nest in deep cavities in old, decaying tree trunks or in deep underground burrows. Woodpeckers and barbets even go so far as to excavate spacious nesting holes in healthy as well as decayed wood. If there is a great scarcity of natural cavities, often because of man's intervention in nature's scheme of things, it is not uncommon for other birds, incapable of excavating a cavity themselves, to take over those made by woodpeckers. Many birds nest in natural holes in earth, clay or sand banks or else simply take over burrows abandoned by mammals or scrape out holes in the ground themselves. Some birds even nest in dark caves.

The commonest building material, used by most birds, is dry vegetable matter – stalks, stems, leaves, roots, cambium fibres, bark. The finest materials are moss, lichen, the downy part of plant seeds and plant wool. This fine material is generally used to line the nest cup in the same way as feathers, hairs, wool and spiders' webs, which are of animal origin. Larger birds often reinforce their nests of branches with mud and turf, the outcome being not only a sturdy but also a heavy structure. Mud is used as the main building material not only by the thrush, which lines the nest cup with a mixture of mud and wood shavings, but also by other birds. For instance swallows, martins and the Rock Nuthatch build their nests by cementing small pieces of mud together, only sometimes reinforcing it with vegetable matter. Their nests may be of diverse shapes – open above or closed, with a simple entrance hole or long entrance tube (Rock Nuthatch). Swifts strengthen their nests with the secretion of their salivary glands which hardens when exposed to air. Swiftlets build their nests only of hardening saliva and affix them to cave walls.

The nesting habits and type of nest a bird builds are an inherited trait. A swallow would never build a nest in a tree resembling a Chaffinch's and vice versa a Chaffinch would never cement a nest together with mud. The Skylark always nests on the ground, the Oriole always in the tops of trees. Despite this, however, circumstances often force birds to deviate from their established customs. Thus the Tawny and Ural Owls nest on the ground when there is a lack of natural nesting cavities, ducks often nest in tree cavities or in old abandoned nests in treetops in areas where floods are frequent. An important role is played also by the geographical location. In central Europe the Peregrine Falcon nests on cliffs, in northern Europe and Siberia in trees and in the tundra on the ground.

Who builds the nest — the male or the female? In the case of polygamous species the task falls to the female, the same being true in most paired, monogamous birds. The male is generally concerned with staking out and guarding the nesting territory. However, in species where both partners appear to build the nest the male often merely accompanies his mate when she collects material or else merely brings the material while she builds the nest alone. The length of time it takes to build the nest depends on many factors — the abundance of nesting material, the weather, time of year, size of the nest and participation of the partners in its construction. In normal circumstances small birds take four to six days to build the nest. The complex, spherical nest of the Penduline Tit takes three to six weeks of work on the part of both mates. The Golden Eagle takes a full two months to build his nest, the Osprey about fourteen days. Migratory birds which arrive at the nesting grounds later than other individuals of the same species build their nests more quickly.

It is commonly believed that the nest is a sort of home for the birds to which they return, at least to roost, even after the breeding period and which they use for nesting several times in succession. In actual fact, however, young birds practically never return to their natal site after they have fledged. Songbirds, in particular, regularly abandon their old nests and build new ones. It is a different case with large birds that build large, elaborate nests, as for example storks. These return to their nests regularly, repairing and adding to them every year. Some large raptors have several nests in their breeding territory, inhabiting a different one every year. Faithfulness to the nest cup is to be found also among small cavity-nesters. There are extremely few instances, however, of birds returning to their nests to roost. As a rule they prefer roosting in the tops of trees and bushes or on the ground. Cavity-nesters, however, are fond of sleeping in cavities. Whereas swallows pass the night amongst reeds, their relatives the martins fly to their nests of mud under eaves as evening draws nigh. Even individuals of the same species may differ in their habits. Whereas flocks of House Sparrows pass the night in their regular roosting trees, other individuals retreat to their untidy nests.

Over the ages some species of birds have completely lost their nest-building instinct. These either fight and seize the nests of other birds or else take over old abandoned nests, behaviour that is very common among birds.

Another striking and common phenomenon among birds is nesting in colonies, which may comprise anything from only a few to a great many paired birds. Colonial nesting apparently affords greater protection for the eggs and young birds against various enemies, both birds and mammals. Few predators are bold enough to face a dense flock of bird parents uttering loud cries and launching a joint attack against the intruder. Another important factor that leads to the formation of colonies is apparently the question of food. Colonies are established either in places where food is plentiful or in isolated locations where there is no other suitable nesting site. Birds nesting in colonies also have nesting territories the same as other birds, but in such cases they are limited to the immediate vicinity of the nest. The instinct for colonial nesting is developed in various degrees. Some birds normally nest individually and only under certain circumstances unite to form nesting colonies. A colony, therefore, may comprise only a small number of nests. Most striking, of course, are colonies that number hundreds and sometimes even thousands of paired birds. Such large colonies are to be found primarily among birds living near fresh water as well as marine birds, though they are not uncommon even among songbirds.

NESTING TERRITORIES

The distribution of nesting pairs of birds is not a matter of chance. Individual species are bound to a specific type of environment that corresponds to their ecological needs, where each has a circumscribed nesting territory. These territories are clearly evident in common songbird populations. The nesting grounds are usually selected by the males, which stake them out before the nesting period and defend them against other males of the same species. The establishment of the territory is proclaimed by song, which prior to breeding has the further function of attracting a mate. The males usually defend their territories in a peaceable manner. The appearance of a rival provokes specific behaviour in the male that is characteristic of the given species and includes various postures, movements, ornamentation of plumage, coloration and vocal expressions. These simple means of expression usually produce the desired result of frightening off the intruder but sometimes they may end in a fight in which blood is shed and feathers fly. Many fights, however, are purely symbolic.

The boundaries of the nesting territory are determined by the location of the nest. In countryside of uniform character the territory is usually circular, though it is often influenced by natural boundaries such as the margin of a forest, vegetation growing on the edge of a lake, or by some body of water. As has already been said, nesting grounds are defended only against others of the same species so that the territory of a large bird, for example — a raptor, owl, or corvid bird — may contain numerous territories of small songsters. Nesting territories vary in size, the area being determined by what food the given birds eat and its abundance. Some birds find ample food in the immediate vicinity of the nest, others have to fly several kilometres. For this reason the nesting territories of raptors, for example, embrace many square kilometres whereas those of small birds cover only several thousand square metres.

In the case of colonial nesters the birds' nesting territory is limited to the immediate vicinity of the nest, sometimes even just the distance a bird sitting on the nest can reach with its beak. Their feeding territories are, of course, many times larger than their nesting grounds. The concept of a nesting territory is markedly developed in the bird realm. It allows for quiet and successful incubation, adequate food and successful rearing of the young. Besides nesting territories birds

also have feeding territories, winter territories and sometimes also mating grounds. The instinct for establishing these territories is developed to various degrees in different species and in some is absent altogether.

BEHAVIOUR

Birds are very agile creatures and their instinctive reactions to a wide variety of situations requiring an immediate response are highly developed. The life of birds is governed by a wide range of instincts which may be divided into several basic groups according to their biological importance: the instinct to seek and ensure an adequate supply of food, the instinct to flee and to attack, the social instincts and the instincts associated with reproduction. Besides these innate instincts birds' behaviour is influenced in slight measure by memory and a simple form of learning based on the principle of 'trial and error', birds tending to remember primarily their bad experiences.

For a certain instinctive behaviour to be provoked the bird must be in a certain state. One example is the behaviour of the Pied Flycatcher, which during the nesting season is stimulated to begin its courtship performance by the sight of a tree cavity, whereas outside the breeding season it exhibits no such reaction. In the said instance the bird seeks an opportunity for a given instinctive behaviour to find expression. It does not look for a place to nest in order to establish a home but to satisfy its instinct, thereupon triggering a further instinct, in this case building a nest.

Instinctive behaviour needs no complex set of stimuli to set it in motion; often just a specific impulse suffices. For example, the reaction of birds to the sight of a raptor in the sky is to flee. Experiments have shown that this reaction can be provoked even by a dummy bird. In the same manner the instinct to feed may be provoked in some songbirds by the mere sight of a dummy representing the open beak and gullet of the young. In other words, this is not a case of mother love but instinctive behaviour which automatically follows in response to certain impulses.

The most striking range of instinctive behaviour is that which is linked with the sexual life of birds, be it during the courtship, mating or nesting period. In most species of birds pairing is preceded by a complex courtship performance which often extends up until after the start of nesting. It takes place on the ground, in the air, in waterfowl on the water's surface, and is sometimes surprising in its complexity. Various aspects come into play at this time – movements, coloration, ornaments of various kinds sported by the male as well as vocal expressions. The manner of courtship is characteristic for each species and plays an important role not only in attracting a mate but also in preventing possible cross-breeding and thus extinction of the species.

That the basic behaviour of birds is innate may be frequently confirmed by examples of birds born in captivity. Even though they are separated from their own kind their notes and calls are the characteristic ones of their species though they have never heard them before. Young weavers on reaching maturity surprise one with their architectural skill which they have had no chance of learning from their elders.

Every related group as well as individual species of birds have a wide range of innate habits and behaviour patterns which are provoked by certain impulses. Study and comparison of the behaviour patterns of individual species, genera, families and even orders yield important clues for determining evolutionary relationships.

In speaking of behaviour mention should be made also of the social ties between partners during the breeding season. The monogamous relationship, in which the paired birds remain together for at least one nesting season, is common among birds. It is not unusual, especially in the larger species, for birds to remain paired for longer periods, sometimes even for life. However, polygamy among birds is likewise not uncommon. For instance the males of gallinaceous birds, ducks and the like have several mates and do not concern themselves with the care of the young. Polyandry, on the other hand, where the female has more than one mate, is found only in a few species.

An important role in the life of birds is played by the voice. Birds are unrivalled in the diversity of their voices and in how often they use them. The vocal organ (the syrinx), located at the branching of the bronchi (not in the larynx as in mammals), is often a complex structure and is most highly developed in the songbirds. The vocal expressions of birds may be divided into two groups. The first are the sounds used by birds throughout the whole year to express contentment or fright, to call attention to themselves, to warn of approaching danger, etc. Song, on the other hand, is only partly innate and otherwise learned by listening to the melodies of other birds. Song is linked with the development of the reproductive organs and hormone production; that is why, apart from the few exceptions, birds sing only during the breeding season. A male's song plays a very important role. It serves as a means of courting the female and arousing sexual excitement and secondly it serves to establish the nesting territory. Almost all songbirds begin their spring song before pairing and nesting. Although song may be viewed as a secondary sexual expression dependent on the activity of the reproductive glands, still to a certain degree it precedes the culmination of the male's sexual activity. Song ceases after the breeding period together with the cessation of activity of the reproductive organs. However, in some species or populations of one species song serves not only to establish the nesting territory and these birds can sing the whole year.

Some birds supplement their meagre vocal range with what might be termed 'instrumental music' – a rattling of the bill or drumming on resonant branches and stumps of branches. In many birds of the woodpecker tribe drumming serves as a kind of substitute for song. Another example is the sound produced by the vibration of the wing or tail feathers. The vocal manifestations of birds are characteristic for the given species and serve as a guide for identifying birds in the wild. Birds very similar to one another in appearance are often more easily identified by the voice than by the plumage.

BIRD MIGRATION

The ability to fly is not limited to the bird world. It is well known that some insects and mammals are also good fliers, but these cannot rival the performance of birds. None can match them in endurance, skill and the ability to fly long distances. Not all birds, however, are dedicated travellers. Some spend their whole life in the neighbourhood of their nesting site, remaining there even outside the breeding season. Such birds are called resident birds. Another group are those that do not migrate but range far afield after the nesting season, for instance if food is scarce or the weather bad. And then there are the birds that migrate long distances, the migratory urge being triggered off by the changing hours of daylight and corresponding alteration in the birds' hormonal activity. Not all lengthy flights, however, are migratory. For example, the young of certain species fly far from the nesting grounds when they fledge (e.g. herons) and this is called pre-migratory movements because it precedes the actual migration. Another example is the so-called invasion which occurs in many species of birds at irregular intervals, usually once in several years. On such occasions entire populations of a given species abandon large areas because of lack of food, water, or for other reasons in order to seek a more favourable environment. The invasions of some birds show a certain regularity corresponding to the cyclical changes in the abundance of their favourite prey, e.g. lemmings. Likewise not classed as migratory are the mass congregations of birds in specific, usually traditional places where they gather to moult. Flying to such spots from far and wide are usually those birds that lose their flight feathers at one time and thus go through a flightless period (ducks).

We have thus divided birds into three groups – resident, vagrant and migratory. It must be pointed out, however, that there is no sharp, clear dividing line between them and deviations from the rule are very frequent. Added to this is the fact that populations inhabiting the northern areas of their breeding range are usually migratory, whereas those inhabiting more southerly areas are resident, which sometimes makes it difficult to place definitely a given species in one of the said groups.

Migratory birds leave their breeding grounds regularly every autumn and return again in spring for a certain period. Birds that breed in the northern hemisphere migrate southwards, sometimes veering to the east or west. The flight to the winter quarters and back takes place along a broad front – only in some places such as sea straits, mountain passes and deep valleys do the birds fly in dense formations thus creating the impression of the existence of narrow migratory routes. Some birds travel during the day, others at night. Most striking of all is the migratory flight of birds travelling in large flocks high up in the sky, whereas flights near the ground or even short-distance journeys from one spot to another often pass unnoticed. However, there are many birds that migrate in this inconspicuous manner. The flightless penguins migrate on the water. Similarly certain waterfowl such as the Brent Goose or grebes sometimes alternate travel in the air with travel on water. Among guillemots the latter method predominates. Travelling 'on foot' (from south to north) has been observed only in the American Coot.

When migrating, birds fly at their optimum speed; in small birds this is 50 to 70 km/hr, and in larger birds it is 60 to 120 km/hr. Worthy of note, however, is the daily performance of the individual species. At one time it was believed that a migratory bird simply took off one day and flew without interruption until it reached its winter quarters. This, however, is quite a rare exception. The average distance covered in a day, which of course does not mean it is repeated daily, is generally less than a hundred kilometres in small birds and 200 to 500 kilometres in larger species. Even so, this is quite impressive and it is not surprising that after such long stretches the birds interrupt their flight to stop and rest, feed and gather strength for the next lap. These intervals may last several hours, days or even weeks. The autumn migration is usually slower than the return flight in spring. Birds often fly to their winter quarters by a roundabout route, without haste, allowing themselves to be coaxed into lengthier stopovers by abundant sources of food. In most instances the return route in spring is shorter so that it does not coincide with the outward route in autumn. It frequently goes directly across the continent whereas the autumn route follows the coastline.

Bird migration is apparently a very old phenomenon from the viewpoint of geology. It probably started because of the differences in climate in the various geographic zones. It is quite probable that originally the territory in which a given species nested was the same as that in which it spent the remaining seasons. External factors or other biological factors brought about the separation of the two areas. It is believed that the reason behind this phenomenon was the hours of daylight in summer, the days being several hours longer in the northern latitudes and thus offering more opportunities for collecting food and feeding the young. As summer drew to a close, however, the birds were forced to abandon their homes by the hostile weather conditions and to seek more suitable winter quarters. These regular movements became established over the ages and apparently became even more permanently set during the Ice Age, when vast expanses of the northern hemisphere were covered by the ice sheet and even the surrounding areas were quite unsuitable for overwintering. It is thus probable that the possibilities of procuring food more easily were the main reason for migration. However, it may have been prompted and influenced by many factors and circumstances and thus to this day it is difficult to find a satisfactory explanation.

Just as difficult as discovering the reasons behind bird migration is explaining how birds navigate over such long distances. Birds are born knowing which main direction to follow when they migrate and their ability to find their way to the distant destinations where they pass the winter, as well as back to their nesting grounds, is governed by instinct. This is confirmed by those species in which the young birds set out alone on the long journey before the adults from whom they might receive guidance. Such is the case with the cuckoo. What sense governs the birds' journey and 'guides' them to their destination has been the subject of much conjecture. Science, however, has rejected several of the theories put forward and come up with a different view. Experiments have shown that some birds, e.g. starlings, navigate by the position of the sun. This does not mean that they fly towards the sun during the noon hours and thus find their way south. Their abilities are far greater. Just as we are able to determine the points of the compass according to the position of the sun with the aid of a watch at any time of the day, so certain birds find their way by means of a built-in clock. In the same way some birds that travel at night navigate by means of the stars, as was substantiated by experiments with birds in planetaria. The method of long-distance orientation is apparently not the same in all birds. This excellent power of navigation, however, is not limited to the migration period, which leads to the assumption that birds have always a special inner sense that enables them to establish their geographical position and find the right direction. Even though many scientists have studied this problem at length it remains in great part unsolved and it will probably be a long time before ornithologists and physiologists find final answers to the questions connected with this puzzle.

The comparatively simple method of bird ringing, which has developed into a vast international project since the beginning of the century, has yielded a great amount of information not only about the location of winter quarters, routes and methods of migration but also about many other aspects of bird biology. Thanks to ringing we know, for instance, the life span of birds, the age composition of populations, how their numbers are naturally regulated and many other details – questions which are only beginning to be studied at greater depth in the other classes of the animal kingdom.

REGULATION OF NUMBERS

The principles whereby nature governs the size of bird populations are truly remarkable. Every species lays a certain average number of eggs during the nesting season. Why some have large clutches or several broods a year depends primarily on the age the given birds can reach and on the mortality of the various age groups. In general it may be said that birds which build their nests in safe (either inaccessible or well-concealed) places and have few enemies lay fewer eggs than those that build easily accessible nests, succumb to various unfavourable circumstances and have many enemies. Birds living in the tropics generally lay fewer eggs than birds inhabiting the temperate zones. In a balanced population the mortality equals the number of successfully fledged birds. Changes in number are the result of the disruption of this equilibrium. Long-term observation of ringed birds of certain species has yielded many interesting details about the make-up of bird populations – about the number of birds in the various age-groups as well as about mortality. The earliest losses occur during incubation and while the young are still in the nest. Thus, for instance, in songbirds that nest in cavities the number of fledglings is only two-thirds the number of eggs laid, in open nests it is about a half and in the case of nidifugous species only a quarter. Further great losses are incurred during the birds' first year. In small birds these are usually seventy to ninety per cent. Older birds, which are hardier and more experienced, have a greater chance of a longer life span than birds in their first year. Experience, resistance and wariness increase with the passing years, but a bird's age has set limits according to the given species. In the wild there is little hope of a bird's reaching this limit. In general, fifty to sixty years is the top limit for large birds and ten to twenty-five years for small birds. Records of greater age (over a hundred years in parrots or vultures in captivity) are the exception.

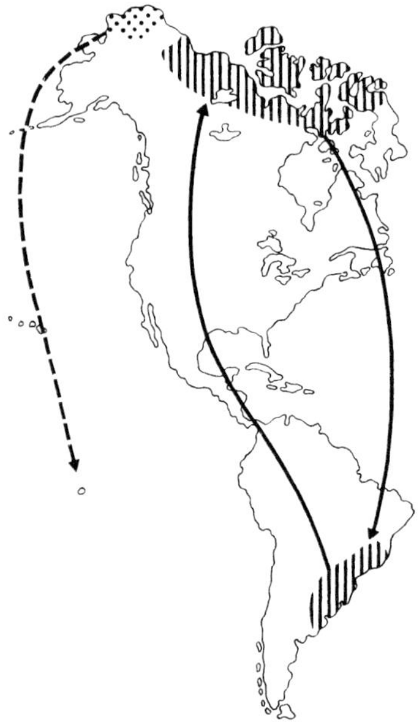

Migration routes of the American Golden Plover

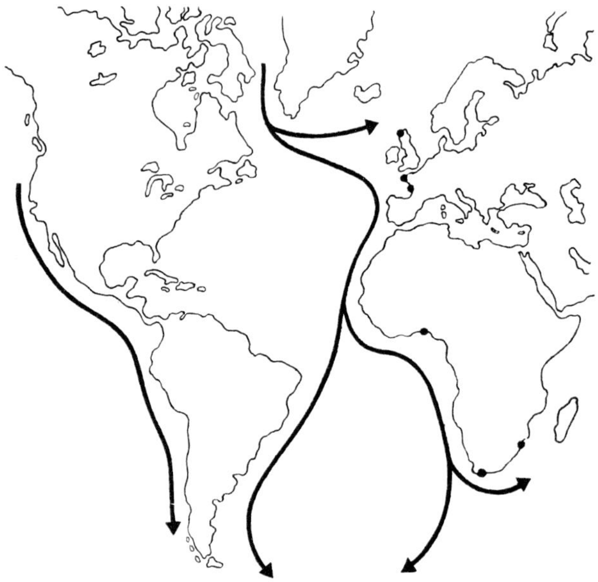

Flight routes of the Arctic Tern from colonies in the North Atlantic and North Pacific Oceans

Man's intervention in nature's scheme of things often disrupts the principles of this balance. He causes the reduction of bird populations by killing birds, destroying their eggs, etc. or indirectly by changing or entirely devastating their environment. In the latter instance much depends on whether the given species is able to adapt itself to the changed conditions or if its numbers begin to decline until the species gradually becomes extinct. It may easily happen that the number of newly reared birds does not keep pace with the mortality rate. The increasing use of various chemicals designed to raise farm production or exterminate pests in farming and forestry, the pollution of the environment with chemical wastes, has today become a greater danger than the direct killing of certain species, even though the disastrous consequences of such actions need not be immediately evident. The numerical decline of various species indicates that all is not well in nature's realm. Often the causes for this decline are not fully appreciated and we do not realize that in the future the same causes may have unpleasant effects even for mankind itself.

During the past four hundred years sixty species of birds have become extinct, most of them exterminated by man. Today many more are on the verge of extinction. Theirs are such small populations that there is very little hope of restoring their numbers. Lest these words sound unduly pessimistic it must be acknowledged that in most cases over a period of years nature is able to make up for the losses incurred by unfavourable circumstances of brief duration, such as sudden hostile weather conditions. As a matter of fact there are several presently known instances of an increase in number in certain species. This is no reason, however, for turning our attention away from the fate of the endangered and declining species, which form the majority.

Besides pictures this book also presents many items of interest about the life of birds. Perhaps it will also serve as an appeal to all to protect birds and prevent the extinction of further species. Birds are not just of importance to man in farming, fruit-growing and forestry and to sportsmen-hunters, but with their delightful presence and song they contribute much towards increasing man's pleasure in the brief periods he spends communing with nature.

Chapter 1 MASTER DIVERS

Penguin jumping from the water onto an ice floe

Many birds are well adapted to life in the water even though, unlike mammals, none spend their entire life in this element. Birds that live in and on the water have a number of characteristic features that distinguish them from species inhabiting dry land. Their body is generally spindle-shaped and excellently protected against the damp and cold by thick, close-packed feathers and a thick layer of subcutaneous fat, especially on the underside where the body is in frequent contact with water. An important role is played by the uropygial gland whose secretion serves to oil the bird's feathers so that they do not absorb water. All good swimmers, and especially divers, have webbed or lobed feet (i.e. flaps of skin on either side of each toe). Only thus are they able to overcome the resistance of the water and attain the required speed when pursuing prey. Birds that dive expertly never fail to rouse our admiration, disappearing as they do in one spot and emerging seconds or even minutes later often at a point that is quite distant.

Most highly skilled in this respect are the penguins, seabirds of the southern hemisphere whose wings have been transformed into powerful paddles. They have no flight feathers and thus none of the seventeen

1

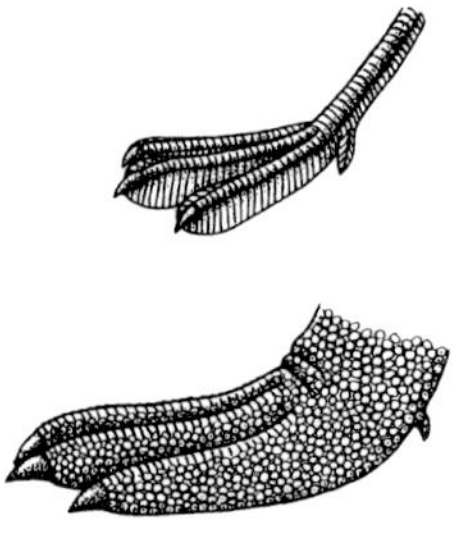

Foot of grebe and penguin

species of penguins can fly. A penguin's plumage is thick, close-fitting and the shape of the feathers reminiscent rather of the fur coat of a mammal than of the ruffled plumage normally associated with birds. The powerful, four-toed, webbed feet are set far back on the body thus enabling the penguin to walk upright and to stand erect, using the short tail with stiff feathers as a support. The diet consists of fish, cephalopods and crustaceans, for which the penguin dives to depths of as much as twenty metres. When diving or swimming penguins use only the wings to propel themselves through the water, the feet serving only for steering and braking. Penguins are gregarious birds, this being most evident during the nesting period when they nest in colonies. Even though civilization is gradually making inroads on penguin populations there are still colonies numbering hundreds of thousands of paired birds. However, oil discharged from ships, shortage of fish and unnecessary disturbance of the birds in their nesting areas are all having a very adverse effect.

The Emperor Penguin (1) is one of the two species inhabiting the Antarctic continent. Large
Aptenodytes forsteri
colonies of Emperor Penguins may be found in barren regions where the temperature drops to minus 50°C. A great measure of adaptation is required to produce offspring in such harsh conditions. The Emperor Penguin does not build a nest and the single egg laid by the female is kept warm by the male in a fold of the belly skin so that it rests not on the ground but on top of the bird's feet. The male is quite skilful at moving about with his burden. In severe winters with freezing winds the male penguins stand huddled in groups to warm themselves while the females range far out to sea for food. For a full two months the males go without food, often losing as much as a third to one half of their body weight. With the aid of a transmitter connected to a thermometer it was discovered that when the body temperature of the adult bird is 38.5° that of the egg does not fall below 28.8°C. Not only the egg but the young, too, are for a long time dependent on the body heat and care of the parents. Not till they are five weeks old do they leave the fold of skin of the mother or father. Both parents take turns caring for their offspring. The young penguin then joins others of its own age group to form a sort of nursery school until it is five months old, when it is able to fend for itself.

The King Penguin (2) nests on islands between 50° and 60° latitude south and is only slightly
Aptenodytes patagonica
different in size and coloration from the Emperor Penguin. It nests once in three years and feeds its offspring a full ten months. In the water it

3

4

moves at a speed of thirty-six kilometres an hour with 120 wing strokes a minute.

The Jackass Penguin (3) is one of the smaller species of penguins. It is a native of the coast and islands of South Africa, its range extending north as far as Natal and Angola. With its wings and feet it scrapes out a shallow nesting hollow in the ground in which it lays two eggs. The young are fed by both parents for a period of three months.
Spheniscus demersus

The Rockhopper Penguin (4) is one of six species of medium-size penguins with a crest. It nests in Tierra del Fuego and other islands of the southern seas. On land it does not walk but moves in hops. Its sharp claws enable it to climb adroitly even up steep rock slopes and cliffs. In late summer and early autumn flocks of these penguins leave their nesting sites and migrate north on the sea currents — riding the waves for a full three to five months.
Eudyptes crestatus

The Peruvian or **Humboldt Penguin** (5) nests in the coastal waters of Peru and Chile where the environmental conditions enable it to breed at any time of the year. In former days, because of its large numbers, it used to be an important producer of guano, along with the cormorants and gannets. Intensive
Spheniscus humboldti

5

6

harvesting of guano, however, has driven this penguin from many of its nesting grounds and has contributed to a substantial decline in its numbers. It nests on rocky shores in caves or in the sparse coastal vegetation.

The Adelie Penguin (6) is one of the few bird inhabitants of the Antarctic. However, there are also colonies of this species on other islands of the southern seas. Either upright or on their bellies, pushing themselves forward with feet and wings, they return in vast numbers, travelling tens of kilometres, to their regular nesting grounds. After the brief courtship period they build simple nests of stones in spots free of snow. Here the female lays two white eggs. She leaves the task of incubation to her partner and goes out to sea for some fourteen days. Then when she takes her turn at incubating the male goes out to hunt and regain the weight he has lost during his two weeks of fasting. Thus there are always a number of penguins travelling to and fro between the sea and the colony.
Pygoscelis adeliae

The Gentoo Penguin (7) is easily recognized by the white band on the crown. Colonies of these medium-size penguins are to be found on certain islands of the southern seas as far north as 50° latitude south. The two eggs are incubated for about five weeks by both partners, each doing a day's stint.
Pygoscelis papua

Other expert diving birds are mostly inhabitants of inland water bodies — lakes, ponds, rivers and their tributaries — and sometimes even ocean

8

lagoons. Divers (order Gaviiformes) have three toes joined by webs. Their nests are located on land near water and in these they lay two pale brown eggs with darker spots. When diving for food the birds use their wings, sometimes descending to depths of fifty metres and when danger threatens remaining underwater as long as ten minutes.

The Red-throated Diver (8) is a typical bird of Scandinavian lakes, partly inhabiting the sparse fringing vegetation which serves to conceal its simple nest located either on the edge or on an island. It feeds mostly on fish which it hunts both by day and night. Night is also the time when the main courtship display takes place, the voice of the birds carrying a great distance. As soon as the lakes begin to freeze over, the Red-throated Divers depart on their southward journey as do other species of divers.

Gavia stellata

The White-billed Diver (9) is the largest of the four known species of divers. Its nesting range in the arctic regions of America and Asia is very limited compared to that of the others. Only during migration does it stray further south.

Gavia adamsii

9

The different stages of the Great Crested Grebe's courtship display

The best diving birds also include the grebes (order Podicipediformes), which closely resemble the divers in appearance. In view of their similarity one would think they were closely related, but their evolution followed separate lines and the relationship is only superficial. One of the most marked external morphological differences is that instead of having webbed feet the grebes have lobed feet — broad membranous flaps on either side of each toe. There are also evident biological differences between the two orders. Whereas divers nest on land grebes build huge floating nests of water vegetation, either anchored to the shallow bottom or else floating freely amidst the vegetation. They nest once a year. The three to eight eggs are white. When leaving the nest the adults often cover the eggs to hide and protect them from enemies. Previous opinions that the heat produced by the rotting nest material serves to incubate the eggs have proved to be false. Grebes feed mainly on aquatic animals. Fish are the mainstay of the diet of the larger species. To keep the fishbones from damaging the lining of the stomach the grebes swallow their own feathers, plucked from their bellies, or occasionally found floating on the water's surface. In the stomach they form a felt-like mass that envelops all undigested parts of the fish and is then regurgitated as a ball. There are nineteen species of grebes distributed throughout the world and like the divers they occur mainly on freshwater lakes and ponds.

The Red-necked Grebe (10) is a native of Eurasia and North America. In its nesting grounds it makes its presence known by its striking laughing note. It keeps to the edges of large reed beds which provide it with concealment and when danger threatens it submerges quietly, staying underwater for more than a minute if necessary.

Podiceps grisegena

10

The Great Crested Grebe (11) occurs in all continents except America and is one of the most attractive inhabitants on bodies of still water. It builds its floating nest in the vegetation bordering the shore either individually or in small colonies. During the mating and nesting season it performs its complex courtship display, consisting of several stages, on the open water. The young have attractively striped plumage and a few hours after hatching are able not only to swim but also to dive when danger threatens. For the first twenty days or so, however, they prefer to ride on the backs of their parents, concealed amongst the feathers with only their heads showing. In this manner they are also flown over the water's surface and sometimes, albeit unwillingly, taken on dives. As the summer draws to a close, European Great Crested Grebes show signs of restlessness and, though during the nesting period they are reluctant to take to the air, at this time they may often be seen winging across the sky. They are clumsy at getting off the ground but once airborne they are rapid fliers. Shortly after they set out on their journey to Africa.

Podiceps cristatus

The Little Grebe (12) is rather inconspicuous, both in size and coloration, and boasts neither a crest nor a collar. It lives a secretive and solitary life amidst the dense vegetation of small ponds and lakes in Europe, Asia, Africa and the Sunda Islands. It reveals its presence in spring by its distinct trilling call.

Podiceps ruficollis

The Black-necked Grebe (13) is considerably smaller than the Great Crested Grebe and is more social, often forming large nesting colonies. Unlike the larger species, which feed mostly on fish, the mainstay of its diet is various invertebrates. It is capable of moving at a speed of four to seven kilometres per hour over the water's surface and, if absolutely necessary will dive to depths of twenty-five to forty metres. Its offspring dive for food by themselves at the early age of seventeen days.

Podiceps caspicus

11

12

13

Chapter 2 THE BEST RUNNERS

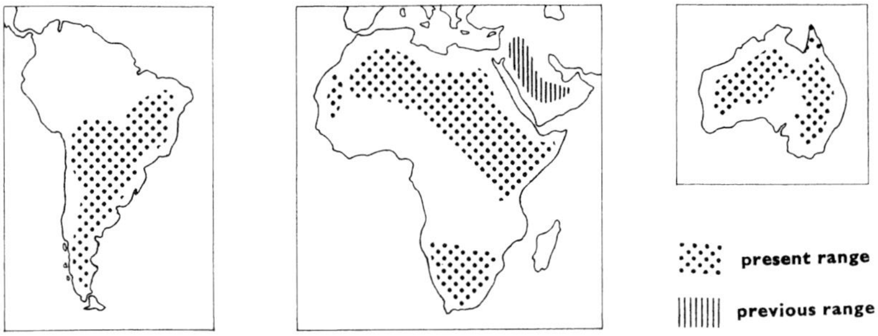

Range of the Rhea (left), Ostrich (centre) and Emu (right)

Most birds take to the air when danger threatens, their escape aided by their agile wings. Escape via land is the route chosen by those birds that have poor powers of flight and that is why the best runners are to be found amongst the flightless ground birds. They belong to four different orders comprising very few species, which have a number of characteristics in common – rudimentary wings, stout legs with fewer toes (except in the case of the Kiwi), a long neck, small head and great body weight. In principle, however, the wing's basic structure is the same as in other birds, which indicates that these flightless birds evolved from aerial forms. The increased body weight and inability to fly were responsible for the reduction of the breast muscles and of the keel on the breastbone.

The Darwin's Rhea *Pterocnemia pennata* (14) is somewhat smaller than the Common Rhea and inhabits the foothills of the Andes.

The Common Rhea *Rhea americana* (15, 18) inhabits the grassy savannas of eastern South America and is often found in the company of herds of deer or semi-wild cattle. One cock

14

15

16

usually has several hens which lay their eggs in one nest — a simple depression made in the ground by the male. The clutch may thus comprise several tens of eggs and it is no wonder that often the male is not able to cover them all when incubating, for it is he who incubates and attends upon the young. Compared with the other running birds the Rhea has quite long wings which it uses for self-defence. During its rapid passage over the ground, when it makes up to 1.5-metre-long leaps, the wings serve as a rudder if it suddenly changes course.

The Ostrich (16) is the largest of living birds, measuring 140 centimetres from the ground to the top of the back, with males attaining a weight of up to 150 kilograms. The structure of the leg indicates that this bird is an excellent runner, what with its strong thigh and tibial muscles and long, powerful feet which have only two toes (the third and fourth). The Ostrich inhabits the African savannas, deserts as well as semi-deserts and is also found in sparse vegetation (bushes and trees). It has great stamina and can run at a speed of fifty kilometres an hour for as long as half an hour. It is capable of a speed of seventy kilometres an hour for short periods. The courtship display of the male during the mating season is a spectacular sight. During this performance he spreads his short wings with their ornamental plumes, circles them rapidly, sits on his feet, inflates and

Struthio camelus

stretches his neck until it is horizontal with his back, then sways his head from side to side until it strikes his flanks, all the while emitting a guttural note. The female lays about fifteen eggs, each weighing about 1.5 kilograms, in a simple depression in the ground. During the day she incubates the eggs while at night the male takes over this task. The young are looked after by both parents.

The Australian Cassowary (17), as all cassowaries, is closely related to the Emu. Even though these solitary birds have stronger legs in relation to their size than has the Ostrich, they are not such fast runners, for they do not live in open country where speed is of vital importance but in the dense forests and jungles of New Guinea and northern Australia. On the other hand, they are better jumpers. It is no great feat for a cassowary, a bird weighing as much as eighty-five kilograms, to leap 1.5 metres into the air from a standing position. When in danger cassowaries defend themselves by kicking with their feet. The female lays three to eight eggs in a simple depression on the ground. They are green, rough textured and weigh over half a kilogram. The clutch does not comprise eggs laid by several hens. They are incubated and the young cared for by the male alone. The diet of cassowaries consists mainly of the fruits of trees which they gather on the ground. Besides this they also eat various seeds, leaves and insects as well as various small vertebrates.

Casuarius casuarius

The Emu (19) is a typical creature of the open plains of Australia, along with the kangaroo. It is mercilessly hunted, however, because it grazes on grasslands and

Dromaius novaehollandiae

17

18

represents serious competition to sheep. The Emu stands up to 1.5 metres tall and weighs fifty-five kilograms. Its strong feet have three toes. The basic wing structure is greatly reduced and the wing itself so small that it is almost entirely hidden beneath the body plumage. In open country the Emu runs at speeds of up to fifty kilometres an hour. During the courting period, which takes place before breeding, groups of birds divide into pairs. The male then scrapes a hollow in the ground somewhere under a bush and lines it with dry vegetation and the female lays seven to twelve green eggs in the nest. Both parents share the duties of incubating and caring for the young. However, the sexes are so alike that it is difficult to tell which does the greater share of the work.

19

20

The Common Kiwi (20) and two other species make up the last order of running birds. On its night-time forages it slowly makes its way through the undergrowth of dense forests using its keen sense of smell to find food. The nostrils open at the very tip of the long bill. During the day it remains hidden in ground cavities or under tree roots, where it also nests. In cases of necessity it can run rapidly for a short distance. The two eggs are incubated and the young cared for by the male. Kiwis are to be found only in New Zealand.

Apteryx australis

The Spotted Tinamou (21) belongs to the order Tinamiformes which includes forty-five species resembling the gallinaceous birds. They inhabit the forests of South America and are closely related to the running birds. They have strong legs, four toes, are excellent runners and clumsy fliers, covering only short distances.

Nothura maculosa

21

Chapter 3 ROAMERS OF THE OPEN SEAS

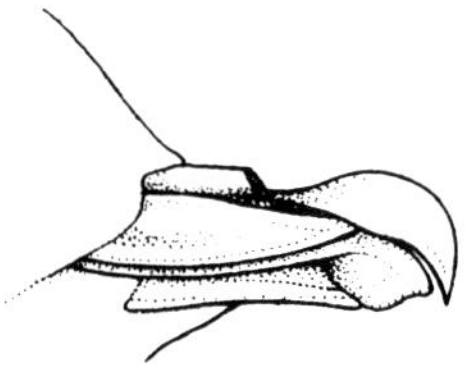

Bill of the Fulmar

Many birds are bound to the sea by their way of life but the mostly highly pelagic are the tubenoses (order Procellariiformes), numbering some one hundred species found in all the seas of the world. About two-thirds live in the southern hemisphere. A truly unique characteristic of these birds is the structure of the bill. It is made up of separate horny plates with the nostrils extending through two horny tubes on the sides of the bill or through a single tube on the top, or culmen. No other birds travel as far from land as do these eternal vagrants with their superb powers of flight. Tubenoses are not noted for their speed but for their endurance in flight. They are capable of remaining aloft for hours without a wing-beat, taking advantage of the wind currents. This is particularly true of the medium and large species, the best known in European waters being the Fulmar, which glides high in the sky only to swoop down to the waves and then up again, with only an occasional wing-stroke. Certain small species, on the other hand, fly so close above the waves that they seem to be touching or running along the water's surface. Only during the nesting period do the tubenoses become terrestrial inhabitants when they come to land to their breeding grounds. With a few exceptions they nest in large colonies, either in the open or in burrows or other underground places. The period of incubation and rearing the young is the longest of any birds. In some large species the single egg is incubated for up to eighty days and the young nestling is not capable of flight until it is about 270 days old. Tubenoses feed on crustaceans, molluscs, fish and refuse found on the water's surface. Their favourite food, however, is squid. Some species are good divers.

The Fulmar (22, 23) Unlike certain species of tubenoses that have greatly declined in number during the past fifty years the Fulmar has been expanding its range for some two hundred years now. It is commonly found on the coasts and islands of the North Atlantic and Pacific, where it often forms colonies with other sea birds. The nest is built on steep rock faces and cliffs. If it happens to be carried far inland by a severe storm then it usually dies for it is unable to find suitable food in freshwater lakes and rivers. Like other tubenoses it does not reach sexual maturity until its sixth year.
Fulmarus glacialis

The Mediterranean or **Cory's Shearwater** (24) inhabits only the high seas and its flight resembles that of the albatrosses – a gliding flight after every five to eight wing-beats. It nests in colonies either in rock crevices or screes in the Mediterranean and certain neighbouring islands in the Atlantic.
Puffinus diomedea

The Leach's Petrel (25) is one of the above-mentioned small species that appear to be running along the water's surface when flying close above the water. It nests on the grassy and rocky islands of the North Atlantic and Pacific oceans, roaming south as far as the equator after the nesting period.
Oceanodroma leucorhoa

The Wandering Albatross (26) is without doubt the largest of all sea birds with its weight of seven to eight kilograms and wingspan of 272 to 322 centimetres. It roams tens of thousands of kilometres from its breeding grounds on the islands of the southern seas, often following lone ships for long periods of time. The nest, scraped together with the bill, is a simple mound of material with a depression in the centre in which the female lays the single egg, weighing almost half a kilogram. These eggs are often collected and eaten by man. The young, fed in the beginning on so-called stomach oil, sometimes attain a weight even greater than that of the parents. The feeding period lasts more than 250 days.
Diomedea exulans

23

24

The Galapagos Albatross (27)
Diomedea irrorata

Colonies of these birds are to be found on one of the Galapagos Islands. The Galapagos Albatross is thus the only species to nest in the tropical zone. Its courtship duties are highly histrionic. The male begins by circling the female with a rocking dance step. After this the two birds swing their heads from side to side and fence with their bills. Now and then one of the two begins clattering its bill like a stork while the other pretends to preen its shoulder feathers. The most striking part of the ceremony is when the bird stands with head erect and sounds a trumpet-like note. Also part of the ceremony is designating the site of the nest and last of all comes a mutual preening of the neck feathers. The nesting period lasts from April till July, after which the birds fly out to sea, frequently being spotted off the shores of Peru and Ecuador.

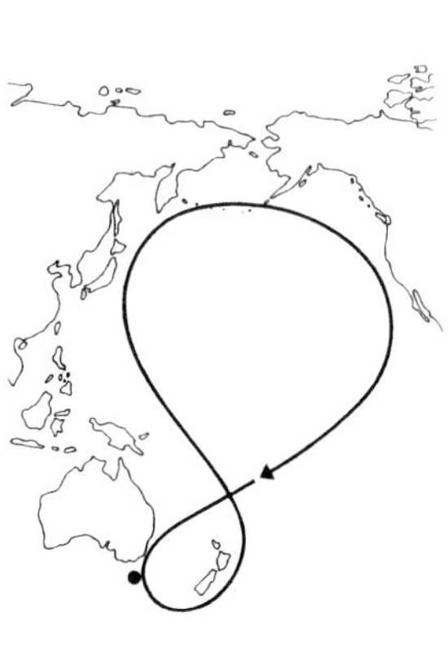

The Short-tailed Shearwater *(Puffinus tenuirostris)* **journeys far into the Pacific from its nesting grounds in New Zealand and Australia**

25

26

27

Chapter 4 THE MOTLEY CREW OF THE PELICANS AND THEIR ALLIES

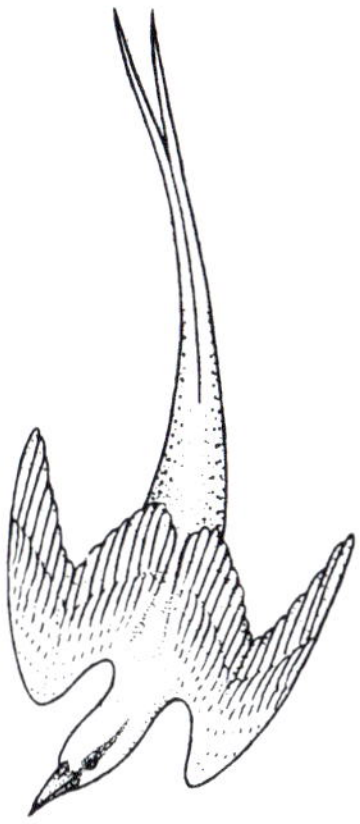

Tropicbird plummeting after prey

The birds of this order (Pelecaniformes) do not embrace a great many species but they exhibit marked differences in appearances, so marked that one would hardly think of classing them together. Their numbers include slender, agile birds the size of a pigeon, medium-sized birds with snake-like beaks, huge clumsy birds with large beaks, expert as well as poor divers, skilful as well as less skilful fliers. Their only common characteristic is that they are all expert fishers. The method by which they capture their prey is reflected in the overall arrangement of the body. Pelicans might be compared to fishermen using a drag-net; tropicbirds and gannets scan the surface waters and on spotting their prey make a steep dive; cormorants and anhingas float on the water's surface and submerge without a sound in pursuit of their prey; frigate-birds catch food from the surface of the sea or pirate it from other birds. All these fish-eating birds have the same foot structure: hind-toe turned forward and all the toes connected by webbing. Their eggs are always a single colour and the young hatch blind and without feathers, needing to be cared for by the parents for a considerable time. The adult birds feed them regurgitated food directly into their bills.

The Red-billed Tropicbird (28) is one of three species of tropicbirds (Phaethontidae) that are gems of the tropical seas. This species inhabits the tropical zone of the Atlantic, Indian and Pacific Oceans.

Phaethon aethereus

28

Pelican feeding its offspring

Tropicbirds resemble pigeons in size and agile flight. They spend most of their lives above the surface of the sea, on which they seldom rest. They catch their food by plunging from a height, disappearing underwater for an instant and then emerging again. They are not active divers for their small and short legs would not be able to overcome the resistance of the water. Unlike other members of the order Pelecaniformes they lay only a single spotted egg. The nests are made in rock crevices, between boulders as well as in shallow burrows which the birds dig themselves. The young have a long fledging period; not till twelve to fifteen weeks of age are they capable of looking after themselves. During the mating season tropicbirds perform magnificent courtship flights.

In every zoo one will find some kind of pelican (Pelecanidae). These are the largest birds of the order Pelecaniformes, some of them weighing as much as twelve kilograms. Their most striking feature is the long, straight bill, only slightly downcurved at the tip, its lower mandible consisting of flexible struts joined at the tip. Between these struts there is a large distensible throat pouch which serves as a dip-net for catching fish. Pelicans are mostly found on large freshwater lakes and large river deltas, even though some are inhabitants of the seashore. They are colonial nesters building their nests in reed beds, on sand or mud banks, islets as well as rocks. These are large but simple structures, being built more solidly only if located in trees. Pelicans are not very well equipped for diving. Their skin and bones are filled with air so that the specific gravity of these large birds is comparatively low.

29

30

The Brown Pelican (30) is an exception amongst pelicans for it is an inhabitant of the sea and
Pelecanus occidentalis the adult plumage is dark. It is found on the coasts of South America and the United States. Flying within reach of the shoreline it seeks its prey. On sighting a fish it plunges down with wings pressed close and neck outstretched, its heavy body striking the water with a tremendous splash that can be heard a great distance away. A moment later it emerges back to front at the same spot with its prey. It nests in large colonies and is an important producer of guano.

The White Pelican (29, 31) conceals its large colonies in thick reed beds. As long as the birds
Pelecanus onocrotalus are incubating their two to four bluish eggs, absolute silence reigns, for the adults are practically voiceless and only now and then do they make a low, grumbling sound. However when the young have hatched, after about a month, and have grown somewhat then the colony presents a different picture for they are very raucous in demanding food from their parents. The young are naked and the bill is not particularly large

31

or long when they hatch. However, as soon as they begin to grow a coat of down the bill begins to increase in size too. At first the adult birds feed the young by regurgitating partly digested fish into the nest, later they do so from bill to bill. The nestlings' whole heads and necks disappear almost entirely inside their parents' bills in their efforts to get at the fish in the throat pouch. The nesting grounds include southeastern Europe, Asia Minor, central Asia and tropical Africa.

The Dalmatian Pelican (32) nests on the freshwater lakes and marshes of southeastern Europe, Asia Minor and central Asia east as far as Mongolia. It is less social than the White Pelican and even though it nests in colonies together with the latter it always keeps apart. Like certain other species of pelicans it some-

Pelecanus crispus

times unites with other birds in flocks to catch fish. In such a case the birds form a long line that slowly advances towards the shore beating the water with their wings and chopping it with their beaks, thus forcing the fish into the shallows where they are more easily scooped up into the distensible pouches. Pelicans fly up into the air with difficulty but once aloft they fly well, a few wing-beats alternating with glides. The typical flight formation of flocks is a slanting line. The birds fly with the head drawn back and bill resting on the front of the neck.

The family of cormorants (Phalacrocoracidae) comprises thirty species. The bill is long and narrow and hooked at the tip. The feet are placed well aft. Cormorants' bones are not pneumatic nor do they have any air sacs under the skin, so that they float deep in the water and dive with remarkable ease. Underwater they propel themselves by paddling with both feet simultaneously; they do not use their wings. The feathers of cormorants are not as water-repellent as those of other water birds and that is why they may often be seen with wings spread out for drying. Cormorants, too, nest in large colonies, sometimes together with other water birds, building their nests in trees and shrubs, on cliffs, between boulders, and in rock crevices, depending on the species. However, the nests of even a given species show marked variation.

The Shag (33) is found on rocky coastal cliffs during the nesting season and is a frequent inhabitant of the breeding cliffs of the northern seas. It nests on the coasts of northern, western and southern Europe, North Africa and the Black Sea. Whole flocks may often be found catching fish in calm sea lagoons.

Phalacrocorax aristotelis

The Guanay Cormorant (34) occurs in great numbers on the islands off the western coast of South America. Not without reason is it called the most valuable bird in the world, for over the years the large colonies deposit thick layers of precious guano. The nesting colonies of Guanay Cormorants represent the greatest concentration of birds in all the world. One such colony – on the twin island of Macabi off the coast of Peru – numbers some ten million pairs of these birds.

Phalacrocorax bougainvillei

32

33

The Double-crested Cormorant (35) is one of the cormorants most widespread in North America, where it nests on the sea coast as well as on freshwater lakes. In waters rich in fish it sometimes congregates in flocks, with some birds forcing the fish towards the shallows whilst others dive for fish in the depths.
Phalacrocorax auritus

The Common Cormorant (36) is found in eastern North America, Europe, Africa, Asia, Australia and New Zealand, with colonies occurring far to the north, e.g. in Iceland, Labrador and Greenland. It inhabits freshwater lakes as well as the sea. Because its large colonies require vast quantities of fish these cormorants inhabit only such bodies of water as will assure them of a sufficient supply. The nesting colonies are sometimes so numerous that there may be as many as ten nests on one tree. These are made of thin twigs and lined with grass, reeds or turf. The clutch consists of three to five blue-green eggs with a chalky surface. The birds begin incubating as soon as the first egg is laid so that the young hatch one after the other and show marked differences in size. The period of incubation is twenty-eight to thirty days, followed by a period of eight weeks when the young are fed by the parents first in the nest and later on nearby branches. The adult birds bring the young partly digested food in their throat pouches, later fresh fish and also water. The cormorant attains sexual maturity in its third year; immatures roam far from their natal colonies even during the nesting period.
Phalacrocorax carbo

Typical position of a cormorant drying its feathers

The Darter (37) inhabits freshwater lakes and rivers in the southern half of Africa, southern Asia and Australia. Together with another American species it forms a family (Anhingidae) closely related to the cormorants. Anhingas have a long, sharp beak that is finely serrated on the edges. A striking feature is their snake-like neck held back in an S in repose and even when the bird is submerged. However, as soon as it comes within striking distance of a fish it shoots the neck out and impales it with its beak as with a harpoon. Small fish are swallowed underwater, larger ones are brought to the
Anhinga rufa

34

35

37

surface, tossed into the air and then swallowed. The nest is built in bushes and trees bordering lakes and rivers or in the middle of swamps.

The Northern Gannet or **Solan Goose** (38) together with eight other species form a separate family (Sulidae). They are typical marine birds nesting in colonies on islands or steep sea cliffs. Two species nest in trees. They dive for their food — fish and cephalopods — from a height of twenty to thirty metres. With wings pressed close they cut through the water, remaining below for a long period. The Northern Gannet nests in the North Atlantic.

Sula bassana

38

39

The Great Frigate-bird (39) of the family of frigate-birds (Fregatidae) nests on the islands of
Fregata minor

the tropic zone in all three world oceans. These remarkably long-winged birds spend most of their time in the air. The short, only slightly webbed feet are not very useful paddles for swimming, not to speak of diving. Frigate-birds seldom alight on the sea, catching their food – various sea creatures, young birds as well as fish – from the surface while remaining airborne. Generally, however, they pursue gannets, gulls, pelicans and cormorants, harassing them until they drop or disgorge their prey which the frigate-birds immediately scoop up in flight. Frigate-birds are extremely aggressive and other birds sometimes suffer quite severe wounds from their long, hooked beaks. That is the reason why many disgorge their food as soon as a frigate-bird comes near. During the courtship season the male's bare throat pouch turns a bright red and he inflates it like a pouter pigeon. Frigate-birds nest in colonies, building their huge nests on the tops of bushes or in trees.

Chapter 5 ALWAYS WITH ELEGANCE

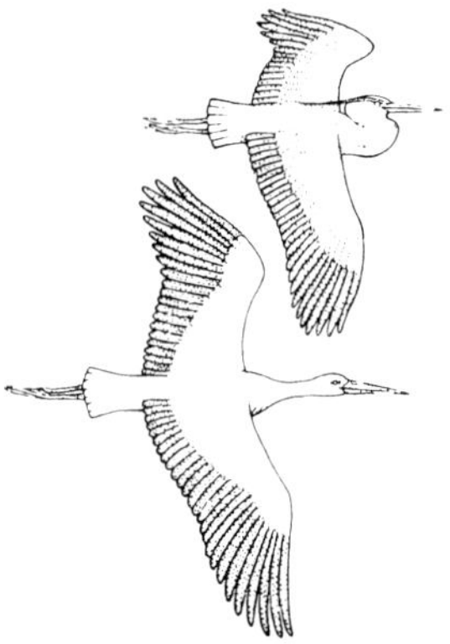

Difference in the flight of the heron and stork

Slender body, long slim neck, soberly coloured plumage, often with long ornamental feathers, graceful movements of the long legs and superb flight are typical characteristics of most wading birds (order Ciconiiformes). It can truly be said of them that they are the most elegant of creatures, a source of aesthetic pleasure and artistic inspiration to man since days of old.

Wading birds have a cosmopolitan distribution throughout the warm and temperate zones of the whole world; the majority of species, however, inhabit the tropical regions. All have long legs with flexible toes often slightly webbed at the base, thus enabling the birds to wade in shallow water and in soft mud. Because all are carnivorous the bill is well adapted for hunting, generally being long and wedge-shaped or otherwise specially adapted, e.g. spoon-shaped (spoonbills) or thin and downcurved with soft tip (ibises). Of the several families the herons (Ardeidae) have the greatest number of species — fifty-nine; the ibises (Threskiornithidae), numbering twenty-eight species, show the greatest variety of shape; the sixteen large species of storks (Ciconiidae) make up a phylogenetically distinct group; and the Boatbill, Shoebill and Hammerhead, systematically important African birds, form monotypic families of their own.

The Common or **Grey Heron** (40, 41) is typical and most common of the herons. It inhabits the Palearctic region from the Iberian Peninsula through all of Europe and the temperate and subtropical parts of Asia to Japan, southern subspecies reaching as far as South Africa and Madagascar. It generally nests in large colonies in tall trees, though sometimes individual pairs will nest by themselves. The clutch consists of four to five blue-green eggs and both parents share the duties of incubating and rearing the young, which they feed mostly small white fish, amphibians and water molluscs, often also fieldmice. Watching a heron foraging for food is a fascinating experience. It either stands motionless or else slowly wades through the shallows until it sights its prey, whereupon it suddenly thrusts out its S-shaped neck, almost always impaling its victim with its sharp bill. The action of thrusting its neck out is made possible by the structure of the twenty neck vertebrae, the sixth being longer than the others.

Ardea cinerea

40 41

42

Certain exotic herons have distinctive methods of hunting, e.g. shading the water's surface with partly raised wings to be able to see to the bottom, and in the case of the Black Heron *(Melanophoyx ardesiaca)* one instance was described of how a bird raised in a marine aquarium learned to use the fish food pellets visitors did not succeed in casting into the water as bait to attract fish for itself.

The Common or **Great Egret** (42), largest of all the white herons, is a magnificent bird with gleaming white feathers and long ornamental plumes which it uses in the manner of a veil during the courtship display. It is a cosmopolitan species nesting not only in southeastern Europe and the warmer parts of Asia as far as Japan but also in Africa, the East Indies, Australia, North and South America.

Casmerodius albus

The Black-crowned Night Heron (43) belongs to the group of smaller, rounder herons with comparatively short legs, but long toes, typical of inhabitants of coastal thickets. It always nests in colonies, numbering as many as several hundred pairs, and is faithful to the same nesting site for years on end. In the crude nest of twigs the female lays four to six blue-green eggs which both parents take turns incubating, both also caring for the young, feeding them small white fish which they regurgitate directly into their

Nycticorax nycticorax

43

bills. During the nesting period the birds keep very quiet during the daytime, their presence being disclosed only occasionally by the cries of the young demanding food. They do most of their foraging in the early morning and at twilight. The young are soon very agile and able to climb amongst the branches, often using their bills to maintain their foothold. The Black-crowned Night Heron inhabits the warmer areas of Europe and all of Asia, Africa as well as North and South America.

The Little Bittern (44), about the size of a dove, is the smallest bird of the heron tribe. The sexes differ markedly in colour. It lives hidden in reed beds where it also builds its typical nest, made of stems laid across each other, in the shallow hollow of which it lays five to six pure white eggs. When danger threatens both the young and adult birds 'freeze' into a reed-like pose that makes

Ixobrychus minutus

44

them invisible amidst the vegetation. It lives in the Palearctic region, Ethiopia and Australia; populations breeding in Europe fly as far as tropical Africa for the winter.

The Eurasian Bittern
Botaurus stellaris

(45) is exclusively an inhabitant of the reedy marshes of the Palearctic region except from when it is also found in South Africa. In Europe it has greatly declined in number in recent years so that its weird, far-sounding cry is heard more and more infrequently in the spring. Unlike all other wading birds only the female incubates and cares for the young. The male often takes two or more mates which build their nests only short distances apart. The nest is always located amidst the dry reeds of the previous year's growth, the bird's brownish-yellow streaked plumage blending perfectly with its surroundings. In early spring the female lays four to five brownish eggs. The bristly, rufous young birds soon disperse amongst the reeds and assume a stiff, reed-like pose. This bittern is rarely seen in flight — only when it wings its way slowly above the reeds from one spot to another. Birds that breed farther north regularly depart to

46

47

warmer areas for the winter though now and then one may come across a cold and hungry bird even in the middle of winter.

The Boatbill Heron (46) is placed in a separate family by itself because of its unusual broad, scoop-like bill, with which it differs from all other members of the heron tribe. However, it greatly resembles the Night Heron in body shape, coloration and behaviour. It remains hidden most of the time in mangrove thickets bordering the rivers from central Mexico to Brazil, nesting by itself or at most in small colonies. Its large black eyes indicate that it is mostly a bird of nocturnal habit.

Cochlearius cochlearius

The Shoebill or **Whale-headed Stork** (47), with its large, boot-like bill is the most grotesque bird of the avian realm. It is found in tropical eastern Africa, mainly on the upper reaches of the Nile. It was known to the Egyptians in ancient times but not for a long time in Europe, for it is an inhabitant of the inaccessible papyrus swamps. It builds its nest amongst vegetation and lays only two white eggs, which become stained a brownish hue by the rotting leaves. It utters a shrill heron-like call but more familiar is the stork-like rattling of the bill with which the birds greet one another.

Balaeniceps rex

The Marabou or **Adjutant Stork** (48) is the best known of the marabou storks (family Ciconiidae), which are unattractive birds but very useful, for they are scavengers that feed mainly on carrion. They inhabit the arid deserts of Asia and Africa, and because of their eating habits are usually held in high esteem and protected by law. With stoic calm they wait for the funeral feast, skilfully grasping their share, their heavy bills a weapon even hyenas and vultures

Leptoptilos crumeniferus

48

eye with respect. They are fond of nesting on solitary trees but have become adapted also to life in the city. Like all storks they communicate by making a hissing sound and rattling their bill.

The Hammerhead (49), because of certain features that set it apart from the other wading birds, mainly its laterally compressed bill and the wide crest plus a number of unusual biological characteristics, is classed in a family by itself. It lives only in the Ethiopian region. One remarkable feature is its queer dance, a kind of act of fraternization, when several birds come together, hop about each other with a clumsy gait, lay their heads back and then

Scopus umbretta

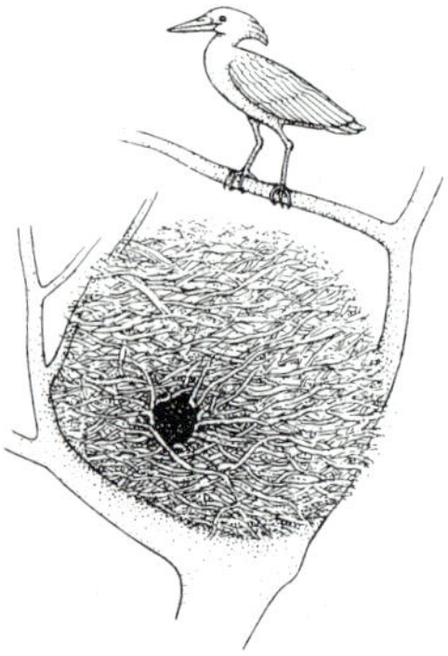

Hammerhead's nest

suddenly fall into a grave stillness. The Hammerhead is noted for its nest, an enormous spherical structure with entrance hole at the side. Though the bird is only about fifty centimetres long the nest measures one-and-a-half to two metres in diameter. It is a sturdy structure made of turf, mud and twigs and is generally located near water in trees or rock crevices; often several nests are placed close together. Both partners apply themselves to the task of building, which sometimes takes several months, but the nest often serves them for a number of years. The birds continually add to it on the outside by plastering on such things as bits of fabric, bones, mud, etc. The clutch consists of three to six white eggs incubated by both partners, who likewise share the task of rearing the young. The Hammerhead hunts crustaceans and small vertebrates in shallows.

The family of ibises (Threskiornithidae) is a group of medium-size birds smaller than storks. Most have long, slightly downcurved sensitive bills with which they probe in the mud for small animals. They fly with their necks straight out in front and are incapable of gliding. When sleeping

49

50

they thrust their bills under the scapular feathers. In general they are gregarious birds and nest in colonies, often together with herons, night herons and cormorants. They breed throughout the warm and tropical zones of the Old and New World excepting Oceania.

The Sacred Ibis (50) figures prominently in certain cults. Five thousand years ago it was venerated by the ancient Egyptians, who made it the symbol of their religion and learning. The god Thoth, scribe of the gods, was depicted as having the head of an ibis. In those times the ibis was a common bird in Egypt, appearing during the spring floods on the Nile which promised a rich harvest. Today it is rarely seen in Egypt, the main area of its distribution being tropical Africa south of the Sahara. The mainstay of the ibis's diet is small animals, the larvae of aquatic insects, and amphibians, though it is also fond of grasshoppers. It is a very social bird, with slow, deliberate movements and buoyant, dignified flight.

Threskiornis aethiopica

The African Wood Ibis (51) is a member of the stork tribe even though its slightly downcurved bill somewhat resembles that of the ibis. The flight and behaviour are practically the same as the White Stork's. It nests colonially in trees and on cliffs in the tropical zone of South Africa.

Ibis ibis

The Saddle-billed Stork (52) is the tallest species of stork standing up to 150 centimetres high. It has an unusually bright-coloured bill – crimson with black band and yellow, saddle-like shield on the top. It lives in tropical Africa where it builds its large nest in the tops of tall trees.

Ephippiorhynchus senegalensis

The Hermit Ibis (53) is a little known species that is becoming extinct. It builds its nest on seaside cliffs as well as in high mountains, sometimes also in ruins. Its

Geronticus eremita

51

52

present range is circumscribed and not known exactly; it has a scattered distribution in North Africa in the Atlas Mountains, by the Red Sea and in Syria. Up until the 16th century it nested in southern Europe in the Alps, and even on cliffs at Salzburg.

53

54

The Jabiru (54) is a New World stork which inhabits South America, its range extending to Mexico. It is almost one-and-a-half times as large as the White Stork and has a tremendous bill, black naked head and neck coloured black with a red ring.
Jabiru mycteria

The Black-necked Stork (55) is found throughout the whole of India, Pakistan and Sri Lanka, but nowhere is it plentiful. It lives in pairs and builds its nest in the tops of tall trees, showing a preference for old, solitary fig-trees.
Xenorhynchus asiaticus

The Scarlet Ibis (56), without doubt the handsomest of the ibises, is an American species found in the southern United States southward as far as Brazil and Venezuela. It nests in trees, often in large colonies numbering several thousand pairs of birds. It occurs in two colour phases: white and scarlet, both frequently nesting in the same colony, generally in mixed pairs.
Eudocimus ruber

The White Stork (57) is a well known and popular bird that since days of old has generally lived in the vicinity of human habitations. In Europe, as far back as the early Middle Ages, it was regarded as a good omen if storks built their nests on a dwelling. Flocks of migrating storks flying through the narrow
Ciconia ciconia

Main migration routes of the White Stork

corridor of Asia Minor roused awe and admiration even in biblical times. Thanks mainly to the banding of birds the White Stork is one of the best studied species of the bird kingdom. It is distributed throughout Europe to the central area of European Russia, with the exception of Italy, France, Britain and Scandinavia; in the southeast large populations inhabit Turkey, Iraq and Iran, and in the south, on the African continent, it nests in Tunisia, Algeria and Morocco. In Europe the White Stork is steadily declining in numbers mainly in lowland areas, apparently as a result of extensive land improvement. The White Stork generally builds its nest on chimneys, rooftops or in trees, adding on to it every year so that old nests are sometimes huge structures measuring up to 250 centimetres in diameter and two metres in height and weighing five hundred to nine hundred kilograms. Almost always they house numerous subtenants, generally sparrows and starlings. The White Stork lays three to six white eggs, though the number of offspring is usually less, generally two to three, for the youngest frequently die. The adult birds are very attentive in their care of the young, shielding them from attack by other storks as well as from inclement weather. During this period they observe a regular daily routine of feeding, at which time the young station themselves in a fan-like arrangement with heads towards the centre of the nest where the adult stork regurgitates the food it brings – mostly frogs and fieldmice. The young follow this by rest and sleep, then careful preening and finally exercises, consisting of stretching, walking about the nest and flapping their wings up and down. The young devote increasing periods of time to learning the rudiments of flying until they are able to remain aloft above the nest for at least a minute. Only then do they venture to abandon the nest and circle in the sky together with their parents. The flight to their winter quarters follows definite routes, which, except for minor details, are known with accuracy as a result of the banding of several thousands of birds. The main migration route of all populations from central and eastern Europe is southeast across the Balkans, Asia Minor and the Arabian peninsula to Africa and the Nile, along which the birds fly to their southernmost destinations. The route of west European populations is in the opposite, southwest direction across the Strait of Gibraltar to Africa.

5

Storks rattling their bills in greeting on the nest

55

56

58

59

The Black Stork (58) is a very shy inhabitant of vast mountain and lowland forests where it
Ciconia nigra
nests in trees or on rocky cliffs. In the quiet shallows of forest pools and streams it hunts fish, amphibians, reptiles and various insects. Breeding habits are much the same as those of the White Stork. The adult birds fly to hunting grounds far from the nest, sometimes tens of kilometres. The Black Stork's range extends from central Europe across the Balkans and Turkey in the south, and in the north from the Baltic countries across the whole of Asia to the Pacific and Sakhalin. A totally isolated Black Stork population exists in the southernmost part of the Iberian Peninsula. Like the White Stork the Black Stork migrates to its African winter quarters in two directions — southeast and southwest — but along a far broader front. It is quite often seen even in Italy and Greece, places that lie outside the main route of the White Stork.

The Roseate Spoonbill (59) is an American spoonbill ranging from the southern United States
Ajaia ajaja
to Argentina. Adult birds are a deep rose-pink, young birds a paler shade at first. Unlike the Eurasian Spoonbill it does not have a crest, but longish feathers on the crop. It inhabits swampy regions and nests in trees, often in mixed colonies with herons and Wood Ibises.

The Eurasian Spoonbill (60), a close relative of the ibises, is distinguished by its long bill,
Platalea leucorodia
flattened from above and broadened at the tip. It forages for food by wading along with the bill partly immersed, swinging its head from side to side and catching the small animals it encounters. Often several birds form a group that wades along in a line. The Spoonbill flies with more rapid wing-beats than the stork or heron, interrupted occasionally by brief moments of gliding flight. The main area of distribution includes south-eastern Europe and the warm parts of Asia; in western Europe the Spoonbill's isolated breeding grounds in Holland and southern Spain are protected by legislation. The nests are built colonially in reeds, also in trees and vegetation on the edges of lakes and ponds. 6

Chapter 6 LONG-LEGGED BEAUTIES

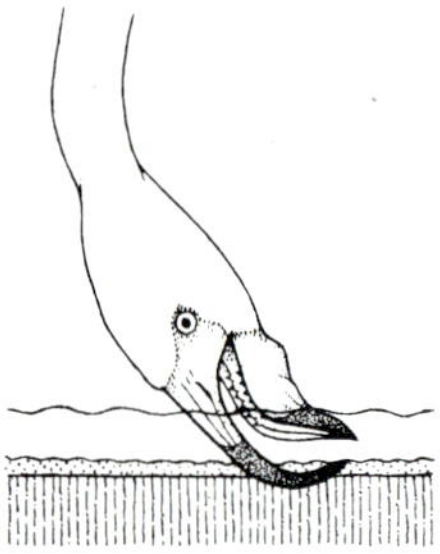

Flamingo's bill

Birds living in an unusual environment and limited to a specialized diet often developed singular shapes and various functional adaptations in the process of evolution. The end result of these changes, however, was not always aesthetically attractive from man's viewpoint. But, richly endowed in this respect are the flamingos, inhabitants of the warm shallow sea lagoons and salt lakes of Europe, Asia, Africa and America. Their beautiful pink plumage and the graceful movements of their slim necks and long legs makes them appear like fairy-tale creatures. A colony of these pink birds in a green lagoon or flying up into the air, their thousands of wings with their black flight feathers contrasting with the azure blue of the sky is an unforgettable sight.

The flamingo's bill is most remarkably adapted for gathering the small animals the bird feeds on from the muddy bottom. It is hooked and the margins are lined with horny lamellae serving to sieve the water pumped from the mouth cavity by the fleshy sensitive tongue. Flamingos feed with the head held upside down, the upper mandible bottom-most and the lower mandible pumping the stirred-up mud into the mouth cavity — a feeding method which is quite unique among birds. Flamingos fly great distances to shallows in search of food — fifty kilometres being quite a normal trek. There are only four living species in the world comprising an old group of birds whose classification remains a problem for they have many features in common with wading birds as well as with waterfowl. That is why systematists generally place them in an independent order.

The Lesser Flamingo
Phoeniconaias minor

(61) is the most plentiful of the small species. It inhabits the salt inland lakes of eastern Africa and northwestern India. It is estimated that some one million pairs nest in Africa and a quarter of a million in India. The filtering equipment of the bill, consisting of thin lamellae, is so perfect that it catches even the most minute diatoms and blue-green algae which are the mainstay of the flamingo's diet. The other two small species, numbering only several thousand birds, are found on lakes in the Andean highlands of Peru, Bolivia and Chile, at elevations even above four

61

62

63

thousand metres. One of these, James's Flamingo, was believed to be extinct until as recently as 1957 when a small number of birds was discovered again in Chile.

The Greater or **Roseate Flamingo** (62, 63), largest of the flamingos, stands about 125 centimetres tall. It has a cosmopolitan distribution; there are several colonies in the Mediterranean and Caspian Sea region, in the Persian Gulf as far as western India, in Africa and in America, where it nests in a number of colonies in the Caribbean area, and in South America from Peru to Tierra del Fuego. Sometimes it changes its nesting site depending on the level of water. There are three distinct geographical races. During the past few years, however, flamingos have been abandoning certain nesting grounds near airline routes for they are extremely nervous of aircraft.

Phoenicopterus ruber

Flamingos live in pairs and nest in large colonies often numbering several thousand pairs. Also unique is their method of building a nest, which is a flattened cone made of mud and bits of vegetation measuring about half-a-metre in diameter at the base and twenty to forty centimetres in height with a shallow depression on the top. Here the female lays her single egg (occasionally two). When incubating, the adult birds sit on top of the nest with their long legs folded so that the 'heel' projects out from under the body beyond the edge of the nest. The young hatch after about a month (28—32 days) covered with thick white down. Their legs are short and the bill straight. They are very active and within a few days of hatching they leave the nest. At first the parents feed them a protein secretion from the crop that is rich in blood and as nourishing as the milk of mammals.

Chapter 7 FAITHFUL MATES

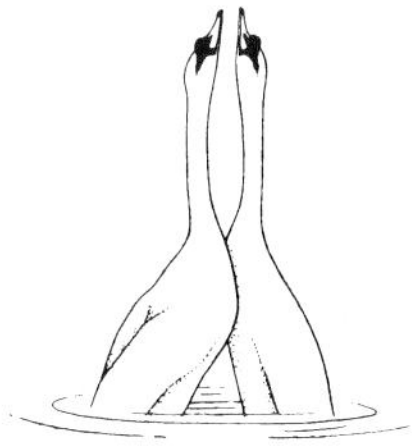

The most spectacular phase of the swan's courtship display is that which follows copulation when the birds face each other on the water, raise themselves up and utter a peculiar deep note

As far as marriage ties are concerned, most species of birds live in pairs, at least during the breeding season. The reason for this is that building the nest, incubating the eggs and obtaining food for the young are very demanding tasks which the female would find impossible to do by herself. Polygamous relationships occur more frequently only amongst nidifugous birds, but in rare instances also amongst other species, such as the Woodcock and Eurasian Bittern. Perhaps the best known examples of faithful mates are the swans and true geese. The paired birds often remain together for life. If one of the partners dies, it is usually only the younger birds that take new mates, older birds often remaining alone for the rest of their days. During the breeding season male swans and geese stand guard over and shield the nest while the female incubates. The males often attack intruders that are larger than themselves, e.g. even man, their chief weapon being their wings, which they beat with great force. When the young have hatched the male joins the family. The family forms a very close unit even though several families may band together for purposes of safety, sometimes remaining together on their journey south and also in their winter quarters up until the time new pairs are formed. Their complex social relationships, adaptability and 'sensible' reactions in widely varied situations rank geese among the most intelligent of all birds. Swans, geese and the widely diversified duck tribe belong to the order of waterfowl (Anseriformes), which is a very distinctive group as regards external features. In body shape and structure of organs these birds are adapted to life on the water. The chief characteristics are the spindle-shaped body, short legs and webbed feet, a thick layer of down insulating the body and close-fitting contour feathers, oiled with the secretion of the uropygial gland, and above all the short, flat bill with small horny plates on the edges and 'nail' at the tip. Most species wade in mud with bill immersed, filtering food from the water, while others use the bill to nip leaves; only in certain specialized species is the bill a long, serrated instrument adapted for catching fish.

Swans need no detailed description. They are found in both the

northern and southern hemispheres. Northern swans are snow-white (adult birds), whereas southern swans are predominantly black. There are only five species of swans in the world.

The Mute Swan (64, 65) is one of the largest and heaviest of all living birds. The male has
Cygnus olor
a wingspan of more than 2.5 metres and weighs up to twenty kilograms. Nevertheless, swans launch into the air with comparative ease, both from the ground and from water. They are an impressive sight both on the water's surface and in flight, with neck straight out in front and wings moving slowly up and down, the wing-beats producing a singing note. The nest, a large mass of dry grasses, is a conical structure, measuring two to three metres across at the base and up to one metre in height, and is built by both partners. The female usually lays five to seven greenish-grey eggs. The young hatch after thirty-five to forty days of practically uninterrupted incubation and are coloured pale grey. The birds do not mature until the fourth year. They are herbivorous birds foraging for food in water as deep as their long necks can reach when they are up-ended. Man has always admired the Mute Swan for the graceful curve of its neck, its carriage while swimming, often with raised wing feathers, and its general appearance, and has usually given it protection. The same cannot be said for the Trumpeter Swan *(Cygnus cygnus buccinator)* of North America whose fate has been cruel and which was practically wiped out by hunters. It was only saved from extinction at the last minute thanks to vigorous protective measures.

The Whooper Swan (66) inhabits a wide area — from Iceland and northern Scandinavia across
Cygnus cygnus
the whole of Siberia and central Asia to Kamchatka and the northern islands of Japan. In the north its breeding range extends as far as the Arctic Circle. In winter Whooper Swans fly to more southerly shores, in Europe mainly the North and Baltic Sea coasts. In severe winters they appear even on lakes in the Alps and in the Mediterranean. Other regular winter quarters are along the Black and Caspian Seas and in the Far East on the waters of eastern China. Unlike the Mute Swan which has formed large populations living in the wild in central and western Europe, where it was originally raised in captivity for ornament, the Whooper Swan remains faithful to its age-old breeding territories in unpopulated areas. In flight as well as on the water it often sounds its melodious trombone-like note.

The Black Swan (67) is the most familiar of the swans that live in the southern hemisphere.
Cygnus atratus
Because of its beauty this native of Australia was brought to Europe where it successfully acclimatized and is now kept not only in captivity

66

67

but also in a semi-wild state. It is distinguished by a remarkably long neck and coral-red bill.

True geese fall into two main groups exhibiting marked differences; the grey geese (genus *Anser*) and the members of the genus *Branta*. Geese of the genus *Anser* are more brown than white, with yellow, orange or bright red legs and bills, which, on the rare occasion, may be black. The fairly large bill is terminated by a broad nail. The *Branta* species, on the other hand, are a striking black colour with various white markings; the small bill terminates in a narrow nail and is always coloured black, as are the legs. True geese are found only in the temperate and boreal zone of the northern hemisphere with the exception of the Hawaiian Goose or Ne-ne, which lives in tropical Hawaii in the Pacific. The greatest numbers of *Anser* and *Branta* species are to be found in the northern tundras of the Old and New World, whence they often migrate far south for the winter, American populations to the southern United States and Mexico, European populations to the Mediterranean region including the coasts of Africa, and eastern populations as far as India and China. The nesting habits and behaviour of all the species are much the same. They form even more permanent relationships than do the swans. All are more or less popular game birds and that is why some species have declined drastically in number, and measures are being taken for their protection.

The Greylag Goose (68, 69), ancestral form of the domestic goose, is one of the geese species that still nest in central Europe, in Bohemia, Moravia and Slovakia, in Hungary on Lake Balaton and the Neusiedler See, in the lake country of eastern Germany and mainly in Poland, where the largest population, numbering some three hundred pairs, breeds. In many nesting grounds the geese have disappeared either because they have been killed off by

Anser anser

hunters or because of changes in their habitat. The breeding season begins in early spring with a colonial courtship performance, with the geese usually laying their first eggs all at the same time. The nest varies in size depending on the location. It is a large structure if placed in reeds on the water and a small one, if built on a dry islet, sometimes consisting of only a few stems and a small mound of down plucked from the goose's belly. The eggs, generally four to seven, are white with a grainy surface and are incubated by the hen alone for twenty-five to thirty days. The goslings are greenish-yellow with dark legs and bills and within a few hours of hatching, after their feathers have dried, are led out by the parents onto the water to safe reed beds. An interesting fact is that during their first few hours the goslings have not yet learned to identify their parents' voice and image so that if they come in contact with man at this time they view him as the parent and follow persistently in his wake. While rearing the young the geese remain well hidden and may be seen only in the early morning or evening when they venture forth to forage for food. At this time the adult birds also moult, shedding all their flight feathers simultaneously so that they are unable to fly for several weeks. Frequently a great many families as well as non-breeding birds from the

69

neighbourhood congregate for the moult in suitable places, usually in large reed beds. Geese spend much of their time on land, where they feed almost exclusively on the leaves of various grasses. Often they 'march' far inland with their goslings, mostly at night, for such tasty morsels as green corn, maize, and the like, flying to water to drink, bathe and whenever danger threatens. When geese are feeding or resting one of their number is constantly on guard, instantly sounding a warning signal when danger threatens. The geese's complex social relationships are expressed by various notes and calls very like those of domestic geese. Quite a number have been recorded, including, for example, soft and contented cackling, the talk of two mates, the call to meeting, the warning cry, the cry of triumph and of course the many notes of courtship. The various vocal expressions are accompanied by typical postures or behaviour. The life of geese has a regular day and night regimen which the birds strictly adhere to. They likewise traditionally observe various habits determined by differing local conditions.

The Bean Goose (70) is the most plentiful of the fourteen species of geese found throughout the world, even though it, too, is declining in number as evidenced by the progressively smaller number of birds migrating to and wintering in Europe. The flocks that numbered several thousand birds in the pre-war years are seemingly now a thing of the past. The Bean Goose breeds in

Anser fabalis

70

71

The welcoming act of the Greylag Goose is strongly reminiscent of the domestic goose of our farmyards

a vast area extending from the eastern coasts of Greenland, Iceland and northern Scandinavia across the whole of northern Asia. There are a number of geographical and ecological races differing mainly in size and in the yellow and black colour on the bill. It is a general rule that the birds of the treeless tundras of the north are smaller and have a shorter neck whereas the races that inhabit the wooded taigas are larger and have a longer neck.

The Snow Goose (71) with its snow-white plumage obviously is an inhabitant of the northernmost
Chen caerulescens
arctic regions. It breeds mainly in the Arctic Ocean along the coasts of Canada and Greenland, on the continent of Asia only in the Chukotski Peninsula and Wrangel island. It has two colour phases – white and grey-blue – which intermingle throughout their range of distribution, both variants occurring in all larger colonies of Snow Geese. It breeds on open ground as soon as the snow begins to melt and reveals bare spots. The breeding colonies are prey to arctic foxes which destroy many of the nests. The main winter quarters of the Snow Goose are the west coast of the United States and Gulf of Mexico; lesser numbers migrate along the eastern coast of Asia to Japan.

The Lesser White-fronted Goose (72) is the smallest member of the genus *Anser*. Adult birds
Anser erythropus
are barely two-thirds as large as the Greylag Goose. They have a white patch on the forehead like the White-fronted Goose but it extends almost on to the crown. This goose breeds in a narrow belt along the Arctic Circle from northern Scandinavia to the Chukotski Peninsula. Only few birds of this species are to be found amongst the flocks of northern geese wintering in western and central Europe for their main winter quarters are farther east, mostly by the Caspian Sea.

The Chinese Goose (73), like the Greylag Goose, has been a domestic bird since olden times.
Cygnopsis cygnoides
It was most probably domesticated in China some three thousand years ago, in other words later than the Greylag. Domestic birds are remarkably

72

73

74

large and heavy, weighing up to twelve kilograms. These geese are always readily identified by the frontal knob at the base of the bill and the wattle on the neck. Wild birds inhabit the central belt in Asia from the middle reaches of the Ob River to Kamchatka and winter in China and Japan.

The Bar-headed Goose (74), a more southerly species distinguished by two black horseshoe-shaped bars on the white head, nests colonially in large numbers on the lakes of central Asia. Its main winter quarters are in northern India. Unlike the other geese it holds its head tilted back with the bill pointing upwards at a slant.

Anser indicus

The White-fronted Goose (75) has a circumpolar distribution and is a typical bird of the treeless tundra. Even though its numbers have greatly declined in the past century because of hunting in its winter quarters it is still one of the most plentiful species for its breeding grounds have as yet remained undis-

Anser albifrons

75

turbed. When migrating it is fond of joining flocks of other geese. It makes its presence known by its frequent laughing two-syllable call that sounds like 'kio-liok'. It spends the winter far to the south, in Europe as far as the eastern Mediterranean, in the Persian Gulf, northern India, central China, Japan and in North America as far as the Gulf of Mexico.

The Hawaiian Goose or **Ne-ne** (76) is the only goose that breeds in the tropical zone of the Pacific on the islands of Hawaii and Maui. At one time a plentiful and tame bird, nesting in the high lava country, it was indiscriminately shot by white hunters during the past century and its nests were ravaged by wild pigs and dogs so that it was on the verge of extinction. All efforts to preserve this bird, at least in captivity, failed. In 1949 there were only about twenty birds living in the wild and seventeen in captivity. Only since 1952, when nine goslings were successfully raised at Peter Scott's famous Slimbridge Wildfowl Farm in England, has the situation taken a turn for the better. Hundreds of birds have been reared within the past few years and reintroduced on several occasions to their native home in the Hawaiian Islands. At present there are already some thousand birds living there in the wild. The Ne-ne has been saved and serves as an example of man's noble efforts to prevent extinction of a species.

Branta sandvicensis

The Brant or **Brent Goose** (77), with circumpolar distribution, is relatively the most widespread of the genus *Branta*, second only to the Canada Goose. It breeds farther north than any other goose – from the 70th parallel to the coast of northern Greenland and the arctic islands round 82° latitude north. It has four geographical subspecies differing mainly by the width of the white neck band and the light or dark coloration of the belly. It is definitely a sea bird feeding on small seaweeds and spending the winter on the coast in more southerly latitudes; it is rarely seen inland, even when migrating. Its one- to three-syllable ringing call that sounds like 'rott-rott-rott' is heard only rarely. In the past few decades the number of wintering birds has been barely a tenth of their previous number as a result of marine pollution, mostly in coastal waters.

Branta bernicla

The Barnacle Goose (78), a not very plentiful species, inhabits the northeastern coast of Greenland, Spitsbergen, Novaya Zemlya and the islands of the Barents Sea

Branta leucopsis

76

77

78

and winters in the North and Baltic Sea area. It nests in colonies on coastal cliffs and islets. Its numbers, too, have declined markedly.

The Red-breasted Goose (79) sports a striking black and white plumage with reddish-brown breast and foreneck and white-rimmed spot on the head. It breeds in the tundra in a circumscribed area on the lower reaches of the River Ob and winters on the coast of the Caspian and Aral Seas. Recently it has become a frequent winter visitor in Rumania (Dobruja).
Branta ruficollis

The Canada Goose (80) is a typical North American goose. There are many geographical races, from the small and short-necked populations of the arctic tundra to the large, long-necked populations of the wooded and lake country of the south. Since the 17th century, if not earlier, it was kept in England in a semi-wild and later wild state and spread from there as far as Sweden. Today flocks of Canada Geese of Swedish origin winter on the coasts of the Baltic Sea.
Branta canadensis

79

80

Chapter 8 NEITHER GEESE NOR DUCKS

The Magpie Goose

Species combining the characteristics of two or more related species, transitional forms so to speak, are not at all uncommon in the bird world. In no other order, however, are they as pronounced as in the order Anseriformes, which includes species that, besides mixed structural features, often exhibit also mixed habits and behaviour. The structural features concerned are generally the shape of the bill, the presence of feathers with a metallic sheen and differences in the size and plumage of the sexes; habits and behaviour include the length of time paired birds and families remain together, the active rearing of the young, the participation of the male in incubating, the social life with its daily regimen, the vocal expressions, defensive reflexes and the like. These factors find expression also in the common names by which the birds are known in various languages, for often they are called geese in some and ducks in others.

The oddest member of the order Anseriformes, and without doubt a very archaic one, is the Magpie Goose *(Anseranas semipalmata)*, which besides possessing features common to geese and ducks also has characteristics reminiscent of the South American screamers and even of fowl-like birds. Its evolution was apparently quite isolated and separate and to this day it has no close relatives. It is found only in Australia and Tasmania and in the southern tip of New Guinea.

The Cape Barren Goose (81) is another relict species of southern Australia. Its appearance clearly indicates its relation to geese but its isolated evolution has led to certain peculiar characteristics, most striking of which is the bill with large yellow-green cere. It pairs for life but is otherwise not a social bird, and is found in the neighbourhood of water but preferring to stay on dry land. It nests on the ground, surprisingly enough during the Australian winter, in June to August; the female sits tightly on the five to six eggs

Cereopsis novaehollandiae

81

82

83

and practically never leaves the nest. It is interesting to note that this habit of breeding in the cold months has been retained even in captivity in European zoos, where it nests from November to January. The male contributes his share to rearing the young, fiercely attacking any intruder that ventures near the nest. This goose, which was almost on the verge of extinction not so long ago, is today protected in its native land and kept with success in many zoos so that its future existence is assured.

The Crested Screamer (82) is a member of the primitive and very old group of screamers of South America. At first glance they show a marked resemblance to fowl-like birds, particularly by the shape of the head and bill and the strong legs and webless feet, but numerous other anatomical features (e.g. the structure of the bones, the layer of down feathers, the uniform distribution of feathers over the body without bare patches) classes them with the waterfowl. They inhabit marshes and the grassy edges of lakes and rivers where they build their shallow nest of grass. Both sexes share the duties of incubating and caring for the young. After the breeding season the birds flock together in large groups. A most distinctive feature is their habit of singing in concert in the evening, with individual flocks sounding their loud, trumpet-like calls.

Chauna torquata

The Magellan or **Upland Goose** (83) is one of a group of five goose-like species inhabiting the colder areas of South America. In appearance it most resembles geese of the genus *Branta* but in behaviour and many other respects it is closer to ducks; for example it moults twice yearly like ducks, the male has conspicuous metallic wing feathers, and the young are speckled and have an eyeband. The male and female have differently coloured plumage, the female being rufous-brown instead of white.

Chloephaga picta

The Andean Goose (84) inhabits the mountain belt of the Andes from Peru and Bolivia to Chile, occurring as far up as the snow-line at elevations above three thousand metres, where it nests in rock crevices. It keeps mainly to dry land but

Chloephaga melanoptera

84

85

leads the young out on the water. Unlike the preceding species the two sexes are alike in colour.

The Coscoroba Swan (85) is another transitional South American type, allied on the one hand to swans and on the other to tree ducks. It resembles the former in size and snow-white coloration, and the tree ducks in posture and shape of the body, head profile, flat bill and similarly structured legs. Last but not least the coloration of the young, which are spotted with a pattern very like the tree ducks, confirms this relationship. The voice is a high four-syllable note that sounds like a toy trumpet. The Coscoroba Swan is found in southern Chile and Argentina as far as Tierra del Fuego and the Falkland Islands.
Coscoroba coscoroba

The White-faced Tree Duck (86), together with seven other species, forms an interesting group of long-legged duck-like birds of upright posture inhabiting the tropical areas of the Old and New World and exhibiting a combination of the features found in ducks, true geese and swans. In habits and behaviour, however, they are most closely allied to the true geese. Entirely unique to these species is the fact that they generally do not form separate geographical subspecies. The sexes are alike in colour and both male and female share the duties of caring for the young, the male even taking his turn at incubating – the only instance of this among waterfowl apart from the Black Swan. The birds pair for life. Another noteworthy characteristic is that they are mostly nocturnal feeders. The note is a pleasant whistling. The White-faced Tree Duck is found on the American continent from Central America to northern Argentina and also in Africa south of the Sahara as far as southern Angola and the Transvaal.
Dendrocygna viduata

86

88

The Black-bellied or **Red-billed Tree Duck** (87) occurs only in America from Texas to northern Argentina. Like other tree ducks it is fond of nesting in tree holes.
Dendrocygna autumnalis

The Australian Shelduck (88) is one of seven species of shelducks, further kinds of transitional types between geese and ducks. The longish body, short legs and metallic wing feathers are a few of the characteristics typical of ducks. Goose-like characteristics are certain forms of behaviour, the slight difference in coloration between the sexes and the male's participation in caring for the young. An interesting trait is the hostility of mated females towards other individuals of the same species. If they do not succeed in chasing away the other bird, be it male or female, they press their mate until he gets rid of the unwanted guest. The Australian Shelduck is a common bird of southern Australia and Tasmania. It usually nests in tree holes.
Tadorna tadornoides

The Ruddy Shelduck (89) is widely distributed throughout central Asia, its range extending as far as the Balkan Peninsula and the Nile delta; isolated populations breed in the southernmost parts of Spain and in North Africa. It is a very adaptable species found not only in the high mountains of Asia but often even in the neighbourhood of human habitations. It inhabits treeless areas near freshwater and saltwater lakes.
Tadorna ferruginea

87

89

Chapter 9 THE LARGE FAMILY OF DUCKS

Main migratory routes of ducks in Europe and Asia

All sorts of water habitats — seacoasts, freshwater and saltwater lakes, ponds, marshes and rivers, at lowland elevations as well as in mountains in all parts of the world, including scattered islands throughout the Pacific and Atlantic — are brightened by the 103 species of widely varied ducks. Different ways of life and different environments gave rise to seven morphologically distinct groups. These include both species with a practically cosmopolitan distribution as well as so-called endemic species with a small range. Species inhabiting the arctic regions often migrate several thousands of kilometres to their distant wintering grounds. The banding of birds has shown that some birds migrate from eastern Siberia all the way to central and western Europe and even to Africa.

Unlike geese and allied species, ducks are more bound to water. Their body is longer, better adapted for swimming, and the short legs, with strongly webbed feet, are placed towards the hind end, this being the reason for the typical waddling gait on dry land. Practically all prefer animal food. Most obtain their food by dabbling and sieving it out of the water through their bills, which are flat with lamellae on the edges. In about fifty per cent of the species the drakes differ from the brownish females in coloration, their breeding plumage sometimes being indescribably beautiful; in many species, especially in the southern hemisphere, however, there are less marked differences between the sexes. Likewise the 'speculum', an iridescent patch on the wings, is not found in all groups. A distinctive characteristic of ducks is the twice yearly moult: a complete moult, including the down feathers, at the end of the breeding season and another, partial moult in the autumn, at which time only the contour feathers are shed. Only the female incubates and cares for the young. Banding as well as experience in captivity have shown that drakes may remain faithful to their mate for several years and not just for a brief period in spring according to widespread belief.

90

91

92

The Mallard or **Wild Duck** (90, 91, 92) is distributed throughout practically the whole of the northern hemisphere. Of all the ducks it is the earliest to begin nesting, often as early as March. It is not particular in its choice of nesting site, locating its nest not only in waterside vegetation but often also in fields and woods far from water, and sometimes in holes in trees. It has even become well adapted to life in city parks and gardens. The clutch generally consists of seven to twelve yellowish or greenish eggs. At the end of the breeding season mallards often congregate in huge flocks on large expanses of water until the frost forces them to depart for more southerly climes. The farthest they migrate is to tropical east Africa and southern Asia. Intensive hunting and the decreasing number of suitable nesting sites due to land improvement are causing a decrease in the number of mallards in some European regions.

Anas platyrhynchos

The Baikal Teal (93) is a smaller duck found in eastern Siberia from the Yenisei River to Kamchatka and south as far as Lake Baikal. It is a common species in its native land, occurring in flocks of many thousands in Japan and southern

Anas formosa

93

China during the winter months, but is a rare visitor to Europe, though other Siberian species are comparatively frequent winter guests here. The drake has a striking facial pattern — green and yellow patches edged with white on a black ground.

The Shoveler (94, 95) is distinguished by a large, spoon-shaped bill, which is a perfect instrument *Anas clypeata* for sieving food, mainly plankton, from the muddy water in which the bird forages. It is found throughout the whole northern hemisphere but nowhere is it as plentiful as other widely distributed species. It is very particular in its choice of location, being partial to shallow and muddy water edges. Its nest, however, is located in drier places in meadows near water. The nine to twelve, smooth, longish eggs are coloured creamy white or pale greenish-grey and are laid from mid-April to May. The eggs of the Shoveler and other ducks of the genus *Anas* are comparatively small in relation to the size of the ducks and have distinctive pointed and blunt ends wherein they differ from diving ducks whose eggs are quite large and symmetrical. Like all ducks the Shoveler lines its nest with a layer of down which keeps the eggs warm for several hours even when the female is away for a time; before leaving the nest she also covers the eggs with down. Early in the autumn the Shoveler departs for its wintering grounds in the Mediterranean, in Africa flying along the Nile as far as the equatorial belt; it also winters in southern Asia and American populations migrate as far as the Caribbean. It is interesting to note that many Shovelers journey to the warm areas of the Atlantic, to England, Holland and Denmark, flying there not only from western Siberia but also from central Europe as banding of birds in Germany and in Bohemia and Slovakia has shown. Young birds sometimes establish their nesting grounds vast distances from their natal site.

The Gadwall (96), a less numerous species, is partial to warmer climates and is found in the *Anas strepera* temperate zone in Europe, Asia and North America. At present its extreme northern boundary is in Iceland. The difference between the sexes is not very marked. Close up, the drake is distinguished by the chestnut-brown wing coverts and vermiculated grey flanks. It is the only

94

95

duck of the genus *Anas* that does not have metallic feathers bordering the white speculum. It is a migratory species, with its chief wintering grounds located in the Mediterranean, along the Nile, in southern Asia and in America by the Gulf of Mexico.

The Chile Wigeon (97) *Anas sibilatrix* has a black head with green metallic gloss, black and white striped mantle and breast and rufous flanks. As in many of the ducks of the genus *Anas* that inhabit the southern hemisphere the sexes are very much alike. The duck is only slightly smaller and slimmer and the feathers of the mantle are shorter and less distinctive than the drake's. It is found in South America from Argentina and central Chile to Tierra del Fuego and the Falkland Islands. The most striking feature of wigeons is their call. The drake's is a three-syllable whistle, the last very loud and long-echoing. The duck usually answers the drake's call simultaneously so that the whistling ends as a harmonious duet. The South American

97

species is vagrant, flying only to the warmer northern areas of southern Brazil, whereas the Eurasian Wigeon *(A. penelope)* often migrates great distances from central Siberia as far as central and western Europe.

Ducks are generally divided into two ecological groups, namely dabbling ducks and diving ducks. Dabbling ducks, represented primarily by the genus *Anas*, are equally at home in water and on land and have the legs located only slightly towards the rear. When swimming their centre of gravity is at the front of the body and therefore the hind end and tail jut above the surface. They forage for food in water by dipping only the head below the surface or up-ending, maintaining their balance with the feet and the front of the body. They rise easily both from the water and from the ground, sometimes even vertically upward. Diving ducks, on the other hand, are adapted mainly to life on water. When swimming their bodies are submerged with the hind end at surface level. Their feet have broader webs and are placed further back and they evidently find walking on land difficult. This, however, is made up for by the ducks' agility in water. They are fond of diving, quite commonly to depths of one to three metres, and obtain their food underwater, usually within about thirty seconds, though if necessary they are capable of remaining submerged for one or two minutes. When rising from the water they often facilitate take-off by running a short distance along the surface. The nests are almost always placed in vegetation on the water, usually in tussocks of grass and also on small dry islets.

The Common Pochard (98), a typical representative of the diving ducks, is commonly distributed throughout central and northeastern Europe and in central Asia to Lake Baikal. In some places, e.g. in central Europe, it has become so widespread that in certain pond areas it is the most numerous duck of all. It breeds later than the Mallard, from mid-April to June, and usually lays six to twelve large, greenish-grey to greenish-brown eggs. The large size of the eggs in proportion to the duck's body, their oval shape with both ends equally rounded and the darker coloration are characteristic features not only of this species but of the eggs of all diving ducks. Ducks often place their eggs in the nests of other ducks or else several females lay their eggs in the same nest. While the ducks are incubating the drakes get together and pass the time solely in 'male' company. Winter is usually spent in the Atlantic area of western Europe, in the Mediterranean and in southern Asia; many birds, mostly drakes, are getting into the habit of passing the winter on rivers in large cities.

Aythya ferina

The Tufted Duck (99, 100) is a Palearctic species ranging from the British Isles and Iceland across all of Asia to Kamchatka and Sakhalin. In western and central Europe its distribution slowly began to spread from the beginning of the

Aythya fuligula

98

99

century and in the post-war years its numbers have increased so rapidly that in some pond areas it has become the dominating species. The ducks are fond of nesting several close together, usually on smaller islets or in vegetation amidst colonies of gulls. The breeding season begins fairly late, usually at the end of May and in June when vegetation is already fairly tall. The clutch is quite large, consisting of eight to fourteen elongate eggs coloured greenish-grey. Often they are laid in the nests of other ducks, one female being capable of incubating as many as twenty eggs, as observation of several such newly hatched broods has shown. In the autumn many birds remain in their nesting territory until the frost arrives, some then moving to rivers and lakes in cities that do not freeze over in winter, often together with mallards, pochards and coots. Banding, however, has shown that most tufted ducks are migratory. Western Europe, especially the British Isles, are popular winter quarters not only for birds from all of eastern Europe and western Siberia (from as far as the River Ob region), but also for many birds from central Europe. More southerly wintering grounds are located in the Mediterranean, eastern populations flying as far as the Persian Gulf, India, China, Japan and the Philippines.

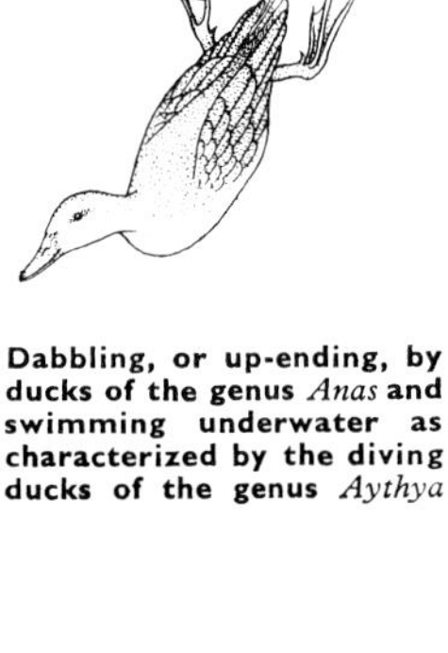

Dabbling, or up-ending, by ducks of the genus *Anas* and swimming underwater as characterized by the diving ducks of the genus *Aythya*

100

The Red-crested Pochard (101, 102), one of the most beautifully coloured of all ducks, is an uncommon species throughout its range. It is a thermophilous duck with a discontinuous distribution in southern, central and eastern Europe, Holland and Denmark being the northernmost limit of its range; its distribution is more continuous in the area beyond the Caspian Sea as far as Mongolia. It is extremely faithful to its nesting site but for years its numbers have remained more or less constant because there are great annual differences in the number of successfully reared ducks. The situation is adversely affected by the fact that there are usually more drakes than ducks and the constant pursuit of ducks by the surplus drakes prevents the formation of pairs. Needless losses are also caused by the ducks' frequent habit of laying eggs in the nests of other ducks which have already started incubating thus resulting in the foreign eggs being abandoned when the young of the sitting duck have hatched. There are marked variations in the times the various pairs breed. Some ducks lay eggs in April, others not until June. The availability of suitable nesting sites – especially dry islets – is generally conducive to earlier breeding. The clutch consists of six to fourteen, pale ochre or yellowish-grey eggs with a matt-glossy surface. The ducklings are easily identified from other species for they are a uniform buff colour. In the autumn northern populations migrate from Europe to the Mediterranean. Large numbers

Netta rufina

101

102

103

of these birds gather on Lake Constance and in southern France at the mouth of the Rhône.

The Common Goldeneye (103) is a northern species inhabiting the taiga belt of Eurasia and North America. It has isolated nesting sites on the lakes of Mecklenburg and Pomerania and the extreme southern point of its occurrence is in south Bohemia, where a small population has been nesting now for the past two decades. A striking feature is their courtship display when the drakes swim around their mates with necks outstretched above the water, every now and then throwing their heads back with bill jutting upward and at the same time kicking up spray with their feet. The Goldeneye nests in tree holes, often as much as ten to twenty metres above the ground, and is also fond of nesting in man-made nestboxes. Newly hatched ducklings are very adept at climbing out of even one-metre-deep holes with the aid of the sharp claws on their feet, after which, encouraged by the calling of the duck below, they make the long jump to the ground.

Bucephala clangula

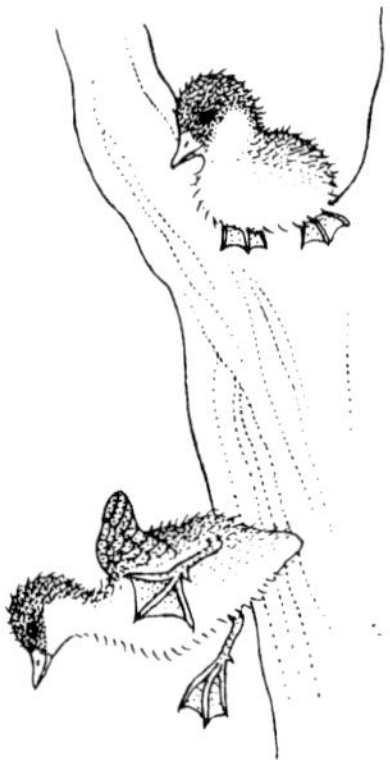

Goldeneye duckling jumping down from the nest

104

105

106

The Wood Duck (104) is a typical North American duck widely distributed in woodlands by
Aix sponsa
stagnant as well as running water. The drakes in their nuptial dress are the most strikingly coloured of all American ducks and for that reason are often kept in captivity for ornament. It can run swiftly along branches, and nests in tree cavities, often several years in a row. For the winter, it flies from Canada and the northern United States to the Gulf of Mexico.

The Goosander (105) together with six other species forms a group of ducks specially adapted
Mergus merganser
for catching fish. All mergansers are excellent divers. Their body is long and spindle-shaped with strong legs and the long, narrow bill has serrated edges and a hooked 'nail' at the tip. They feed mostly on small fish for which they dive to depths of several metres, remaining underwater for as long as two minutes. The Goosander is widespread throughout the northern parts of the Holarctic, but in some places, particularly in the mountainous regions of Asia, it nests even at more southerly latitudes. In Europe there exists an isolated population in the Alps. The Goosander nests in tree holes, quite often far from water. When very small, the young ride on the back of the duck, but unlike young grebes, which snuggle in the down feathers under their parents' wings, the young Goosanders perch on the back. The birds winter in more southerly locations, the extreme southern limit in Europe being the Mediterranean and Black Sea, in America the Gulf of Mexico. Groups of mergansers generally remain apart from other ducks and occur in the same sites every year.

The Velvet Scoter (106) and two other species of this genus are birds of the north, which, apart
Melanitta fusca
from the breeding season, are generally to be found out on the sea. They are dark, plump birds with broad bills that have a protuberance at the base. The diet consists almost solely of bivalve molluscs and crustaceans, for which the Scoter dives to depths of as much as fifteen to twenty metres, a depth attained by no other birds of this order. When diving they use their wings. The Velvet Scoter breeds in the colder regions of Europe and western Asia, converging in winter on the southern coasts of the Baltic and North Sea and along the Atlantic coast to the Bay of Biscay. They are less often seen inland, usually only ducks or young birds, which migrate farther south than the drakes. The latter abandon the seacoast only in particularly harsh winters.

107

The White-headed Stifftail (107) is a member of a group of six thermophilous duck species of both the Old and New World, characterized by long, stiff tail feathers which are held upwards at a slant when swimming. The White-headed Stifftail has a disrupted range in southern Europe and the Balkans; its distribution is more continuous in the Caspian-Aral region up to the upper Yenisei River. It is interesting to note that the eggs are laid in a nest of decaying vegetation which as it rots generates heat that keeps the eggs warm enough for the incubating duck to leave the nest for longer periods than she otherwise could. The eggs are unusually large in proportion to the bird's body and a full clutch of six to twelve eggs weighs two to three times as much as the duck. Also very interesting is the male's courtship display. The drake circles on the water with tail upright so that the white under tail coverts are continually turned towards the duck. He furthermore thrusts his bill sharply against his neck until the air escaping from the neck feathers forms bubbles on the water's surface.

Oxyura leucocephala

Chapter 10 BIRDS OF PREY

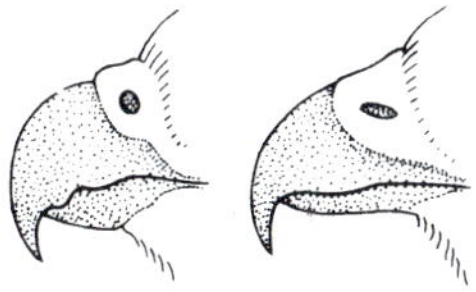

Difference in the shape of the bill of falcons (left) and eagles (right)

Man attributes to birds of prey many of his own characteristics – some justifiably so, others not. Cruelty, ruthlessness and cunning, however, are definitely not among them. The birds in this chapter have been shaped and adapted by nature to hunt prey, to kill weak and sick individuals incapable of survival; and a daring, fighting spirit, speed and skill are natural prerequisites of successful hunting. Not all raptors acquire their food by hunting, however. Some have been allotted the task of getting rid of the carcasses of other animals. Morphologically, raptors are a comparatively uniform group. They all have a relatively short, hooked beak, with sharp edges, strong powerful feet and long claws. The birds of prey (order Falconiformes) comprise 291 different species, all good fliers. Those that seek their prey on the ground or feed on carrion are characterized by slow, enduring flight, others are agile fliers quick in pursuing and catching their prey on the wing, whereas still others are specialists that swoop from the sky into water after their prey, or birds that hunt insects, snakes, turtles, molluscs or ones that are vegetarians feeding on fruit. Most raptors feed on live vertebrates, regurgitating the undigested parts such as hairs and feathers in the form of pellets some sixteen to eighteen hours after their meal; unlike with owls the regurgitated particles in this case contain comparatively few bones. Birds of prey are diurnal and have keen eyesight. Their nests are built in trees, on cliffs, on the ground, in reed beds, in tree holes or in rock crevices. They are nidicolous birds that care for their young a long time in the nest as well as long after fledging. Many raptors are rare species in grave danger of extermination and that is why in recent years efforts to provide adequate protection for their preservation have been winning increasing support.

The American King Vulture (108) resembling the Old World vultures, is the most strikingly coloured of the New World vultures. A lowland inhabitant, it is found in the forests and pampas of Central America south as far as southern Brazil. It feeds on carrion, which it locates (one of the few birds to do so) with its sense of smell. The nest is built in trees.

Sarcoramphus papa

108

109

The Andean Condor (109) is the largest of all raptorial birds (weight: 11 kg, wingspan: 3 metres). *Vultur gryphus* It lives in the high mountains of South America, where its powers of flight enable it to climb to heights unattainable by other birds. It feeds on carrion, rarely attacking live prey. The Condor's hunting territory is not limited only to the high mountains. It often leaves the mountain ridges for the seacoast where it is easier to find food in the form of dead fish, mammals and birds cast up by the sea. It is sometimes seen even in colonies of marine birds where it steals eggs as well as young from the nests. The nest is built on cliffs and the brood consists of two birds at the most.

The Turkey Vulture (110) is the commonest of the New World vultures, ranging from southern *Cathartes aura* Canada to the Falklands. It measures 73 centimetres and has black plum-

110

111

age and a bare crimson head. It may be seen circling above the countryside, either singly or in large flocks. It feeds not only on carrion but also various other refuse and therefore is not persecuted by man. It seeks food by flying close above the ground, locating it with its sense of sight as well as sense of smell.

The Honey Buzzard (111) is one of the food specialists amongst the raptorial birds, its favourite tit-bit being the larvae of wasps and bees. It also feeds on other insects, sometimes captures a small invertebrate and is not averse to eating even fruit. Because it is a comparatively large bird requiring quite a number of larvae and insects to supply its needs the Honey Buzzard roams over a fairly large territory, about 3.5 kilometres within range of its nest. This is built in a concealed spot high up in a tree and the edge is often decorated with green branches. The two to three eggs are incubated for thirty to forty days and about six weeks later the young leave the nest to roam the neighbourhood, returning, however, to spend the night. Duties are not divided between the partners as is often the case with other raptors. The Honey Buzzard is protected from insect stings by a thick coat of feathers covering even the part of the head between the eye and the base of the beak where other birds have only bristle feathers. The Honey Buzzard is a common inhabitant of Europe and northern Asia but being a migrant it winters in Africa and southern Asia.
Pernis apivorus

The Black Kite (112) is a raptor with shallowly forked tail widely distributed from western Europe to Japan and from southern Scandinavia and Siberia to South Africa, Malaysia and Australia. It feeds on all kinds of carrion, small vertebrates and insects and often litters its nest with all sorts of refuse — bits of paper, cloth, leather, rubber.
Milvus migrans

112

Range of distribution of the Andean Condor

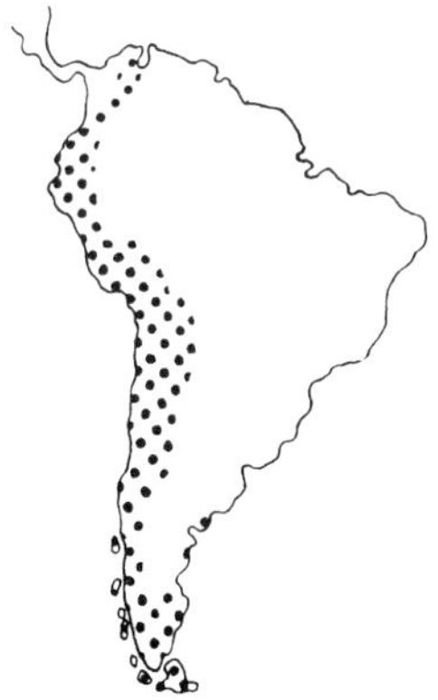

The Secretary Bird (113) differs markedly both in appearance and way of life from the other
Sagittarius serpentarius birds of prey and is therefore placed in a separate family (Sagittariidae) by itself. Its most striking features are the cursorial legs with short toes resembling those of the bustard, and as these indicate it hunts its prey mostly on foot, using its wings only in the case of dire necessity. In the African savannas south of the Sahara it hunts small vertebrates, being particularly fond of snakes. The nest is built in trees or bushes.

The Goshawk (114) was once a common bird of prey in Europe's woodlands. It is perhaps still
Accipiter gentilis fairly common in North America and northwest Africa but in Europe its numbers are rapidly declining. It builds its nest in tall trees, often availing itself of the abandoned nests of other raptors. The male shares the task of incubating the three to four eggs, but to a lesser extent than his partner. The young are cared for by the female alone while the male procures and brings her food. Goshawks do not breed until their third year, at which age they already have transverse bars on the chest and belly. Birds spotted longitudinally — as shown in the picture — are known to nest only in rare instances.

The Sparrowhawk (115) is very similar to the Goshawk but smaller. It lives in Eurasia and
Accipiter nisus Africa. During the nesting period the male keeps the hen provided with food, usually giving it to her outside the nest. She then tears it into pieces before giving it to the young. Sparrowhawks have been becoming scarce in recent years. They feed mainly on small birds.

The Lesser Spotted Eagle (116) is one of the commoner species of medium-size eagles. It is
Aquila pomarina found in central and eastern Europe and in southern Asia and the Middle East. In the spring, like most raptors, the male performs magnificent courtship flights. It is a migrant that returns to the same nesting grounds for years on end. The female lays two eggs, but the birds usually rear only one nestling. The younger dies shortly after hatching because the

113

114

115

Goshawk attacking its prey

elder is more aggressive and thus gets the greater share of the food. It remains in the nest a full eight weeks before leaving it.

The Imperial Eagle (117) is only somewhat smaller than the Golden Eagle. It favours open
Aquila heliaca
country where deciduous woods alternate with flat spreading expanses. It nests in the Iberian Peninsula, southeastern Europe and as far as northwestern India and Mongolia. The bulky nest of twigs, almost always edged with green branches, is built in tall trees. The clutch generally comprises two eggs. This eagle hunts medium-large mammals and birds but is not averse to feeding on snakes as well.

The Golden Eagle (118) is a greater hunter than the Imperial Eagle as evidenced by its more
Aquila chrysaëtos
powerful beak and longer talons (the picture shows a young bird). In Europe it used to nest also at lower altitudes, on cliffs as well as in trees. However, it was so indiscriminately hunted by man that in some areas it has disappeared altogether or else been forced to retreat to the high mountains. It is a huge bird with a wingspan of some two metres. As a rule it hunts its prey by flying close above the ground and surprising its victim by striking without warning. Sometimes it attacks by plummeting from the sky, seizing its prey by the head or back with its powerful talons and killing it with sharp thrusts of its beak, flapping its wings up and down as it does so. The struggle is sometimes a lengthy one. The Golden Eagle lays only one or two eggs. The young remain in the nest a long time, about three months, before they learn to fly. There is usually only one fledgling. Considering that these birds do not reach maturity until the age of five years it is no wonder that such a small brood is unable to make up for the losses caused by sportsmen. The Golden Eagle inhabits a large part of Europe, Asia and North America and still nests in the mountains of North Africa.

The Rough-legged Hawk or **Buzzard** (119) is easily distinguished from the Common Buzzard
Buteo lagopus
even in flight by the white area at the base of the tail and from close up by the feathered feet. It is a typical inhabitant of the Old and New World tundras. Its numbers depend on the abundance of lemmings which are the mainstay of its diet. In winter, when food is scarce because of the thick layer of snow, whole populations of these birds fly south. If there are not enough mice and fieldmice or if the ground is covered by a thick snow blanket the birds may even be responsible for loss of wild game.

6

117

118

119

120

The Common Buzzard (120) is somewhat smaller than the last species and is a common woodland raptor. It hunts small harmful rodents, sometimes also other vertebrates, in open ground and is a very beneficial bird to farmers. The nest is built in trees and the two to six eggs, spotted reddish brown, are laid in April. When they hatch, after twenty-eight to thirty-one days, the young are covered with fine white down feathers; they are fed by the adult birds in the nest for six to seven weeks. The Buzzard nests in the temperate zone of Eurasia.
Buteo buteo

The Bald Eagle (121) of North America is a close relative of the White-tailed Sea Eagle. It nests near large lakes or the seacoast, where it hunts fish or water birds. In former days its numbers were decimated by indiscriminate shooting and to this has now been added the damage caused by pesticides used to exterminate insects and rodents. Traces of these poisonous substances make their way indirectly to the birds' eggs causing hatching failure. Only about one third of nesting birds rear one, occasionally two young.
Haliaeëtus leucocephalus

The Eurasian Black Vulture (122), a large carrion-feeding raptor with wingspan of up to 2.5 metres, nests in trees and on rocks in rugged mountains. In Europe it breeds only on the Iberian Peninsula, in the Balkans and in the Crimea, also in the mountains of northwest Africa and throughout much of Asia. It lays only one egg, the young bird being fed in the nest for about $3\frac{1}{2}$ months. When carrion is scarce, the Eurasian Black Vulture also hunts live prey — marmots and other rodents, domestic animals, snakes and turtles. It often chases off other raptors when attacking its prey.
Aegypius monachus

The African White-backed Vulture (123) is still a common sight above the savannas of Africa where it soars keeping a sharp eye on its companions circling in the
Pseudogyps africanus

121

122

123

124

125

distance. As soon as one of these birds discovers a carcass and alights it is immediately joined by the rest. Before long the bloody feast, in which the vultures are joined by hyenas and jackals, is over and the foul-smelling carcass is no more.

The Lammergeier or **Bearded Vulture** (124) is easily identified in flight from the other large birds of prey by the long, wedge-shaped tail and long, narrow, pointed wings. At the base of the bill it has a striking black beard of bristle feathers. It, too, feeds on carrion, being particularly fond of bones. Though it is able to swallow unbelievably large pieces very large bones sometimes present a problem. To be able to eat them it soars high up in the air and shatters them by dropping them on the rocks. In like manner it shatters the shells of large turtles. The Lammergeier is an inhabitant of the mountains. In Europe it is found only in small numbers in the Balkans and Iberian Peninsula and it has a discontinuous distribution from East to South Africa and in central Asia. The nest is built on cliffs. The two eggs are incubated for fifty-five to fifty-eight days but usually only one young bird is reared by the adults, which feed it for a full three months.

Gypaëtus barbatus

The Bateleur Eagle (125) belongs to a different subfamily than vultures. It has long, narrow wings and an extremely short tail, creating a distinctive flight silhouette

Terathopius ecaudatus

126

as it performs its daring aerobatics above the savannas of southern Africa. It builds its nest in trees or on termite hills, and feeds on small vertebrates and carrion.

The White-tailed Sea Eagle (126) is a close relative of the North American Bald Eagle. Both have an unfeathered tarsus. It lives in the vicinity of large lakes, rivers and the seacoast where it finds an abundant supply of water birds and fish, which form the bulk of its diet. Occasionally it will also catch a mammal, and in case of dire need will even eat carrion. It is a large and powerful raptor with a wingspan of some 2½ metres, but is not a particularly skilled hunter. In wooded lowlands it builds its huge nest in old trees, on the seacoast it nests on cliffs and in the tundra even on the ground. Like other large raptors it has several nests in its breeding territory, using a different one each year and adding to it every time. Over a number of years such structures thus acquire impressive dimensions. The two to three white eggs are laid in February or early March and both parents take turns incubating, though the female bears the brunt of the task. The young hatch after thirty-five to forty days and remain in the nest a further ten weeks. Soon after that they take to the air and roam far from their nests, leading this sort of life until they reach maturity, i.e. until they are four to five years old. In Europe the White-tailed Sea Eagle has retained its foothold only in Norway. Otherwise it has a discontinuous distribution throughout the Palearctic region east of the River Elbe. In most countries it is a protected species and it is only thanks to the strict conservationist measures in its breeding grounds that

Haliaeëtus albicilla

one may now and then catch a glimpse of its flight silhouette against the blue sky.

The Harpy Eagle (127) Much has been written about the cruelty, bloodthirstiness and strength
Harpia harpyja
of this huge raptor of the rain forests of Central and South America. In relation to its size it has the strongest legs, toes and talons of all the raptors and so can go after monkeys and sloths in the treetops. The sloth is an animal of no small size and with its hooked claws can keep a very firm grip on branches. To pry it loose and kill it requires truly great strength. The Harpy Eagle is also known to capture ground mammals and domestic animals and monkeys flee in panic at the sight of this raptor. It is found in the neighbourhood of large rivers and lays its eggs in a huge nest in trees.

The Montagu's Harrier (128) is one of seventeen species of harriers (subfamily Circinae) –
Circus pygargus
slender, long-tailed and long-legged raptors living in open country. All have owl-like facial discs. When hunting Montagu's Harrier flies with a rocking motion close above fields, meadows and steppes. It inhabits a great part of the Palearctic region from Europe to western Siberia but its range does not extend far north. In spring it performs its beautiful courtship flight above the nesting territory, incorporating many acrobatic feats. The simple nest of twigs, shoots and stems is built on the ground amidst a tangle of grass and rushes. The clutch comprises four to five white eggs spotted light brown. Throughout the entire period before the young fledge the male brings food to both his mate and offspring. The harrier's weak talons are not made to catch larger prey and so it feeds on small rodents, frogs, lizards and small birds. The hunting territory embraces a circle ranging four kilometres from the nest. Montagu's Harrier is a migratory species that winters in southern Asia and Africa. It shows marked sexual dimorphism, as do other species of harrier. The male is pale grey with black wing tips and black band across the wings, the female brown with yellow breast and belly spotted longitudinally. The young are coloured onion-brown on the underside.

127

128

The Marsh Harrier (129) during the nesting period is found only in the neighbourhood of water — ponds, lakes and marshes — where it may be seen flying close above the water, hunting small mammals, birds and frogs and where it builds its nest, a pile of dry water vegetation and twigs, amidst the large beds of reeds and rushes. In May it lays three to six eggs, the female beginning to incubate as soon as the first is laid so that there is a marked difference in the size of the offspring; the youngest is often slow to grow and dies. The female rarely leaves the nest for, throughout the entire period the male keeps her supplied with food. Often she flies out to meet him and takes the food while still on the wing. In late August or September the birds leave their nesting grounds and fly south. The breeding territory includes the entire temperate zone of Eurasia.

Circus aeruginosus

129

130

The Eurasian Kestrel (130) is the most common and widespread of the falcon family (Falconidae)
Falco tinnunculus
in Europe, Asia and North Africa. It hunts in open country, often hovering in one spot on the lookout for prey, then suddenly diving to the ground and flying up again with its victim — a mouse or fieldmouse — to a nearby tree, post or its nest, placed in a treetop at the edge of a forest or on a rock. The five to six eggs are laid in April or May and throughout the period of incubation the male brings his mate food as do most raptors of the falcon family. The Kestrel does not occur only in lowlands; it is often found also high up in the mountains where it nests on rocks, hunting various mountain rodents. Kestrels nesting in the north are migrants, those that breed in the southern part of their range are resident.

The Osprey (131) — on the nest — is classed in a separate family (Pandionidae) by itself because
Pandion haliaetus
of its distinctive body structure, wherein it differs from the other raptors.

It is well-formed for hunting live fish. The legs are short and the outer toe is reversible so that the slippery fish can be gripped with two toes in front and two behind and is prevented from slipping from the bird's grasp by short stiff spikes on the pads on the undersurface of the foot. The Osprey is found near water practically throughout the whole world. It generally nests in trees.

The Southern Caracara (132) is a South American member of the falcon family which, with its long, strong legs moves with agility on the ground. Its diet is a varied one, ranging from small vertebrates to carrion. In flight it harasses water birds until they regurgitate the fish they have caught, which it then eats.
Polyborus plancus

The American Sparrow Hawk (133) is a typical representative of the true falcons of America. Its range reaches from the edge of the tundra up to Tierra del Fuego. It is not a shy bird and may often be seen perching on telegraph poles or trees beside highways. It feeds chiefly on insects.
Falco sparverius

The Gyrfalcon – white Greenland race – (134) is a noble and powerful falcon, stronger and more daring than the Peregrine Falcon. In the Middle Ages it was highly prized as a hunting bird. It is an inhabitant of the far north.
Falco rusticolus

131

A Peregrine Falcon attacking its prey

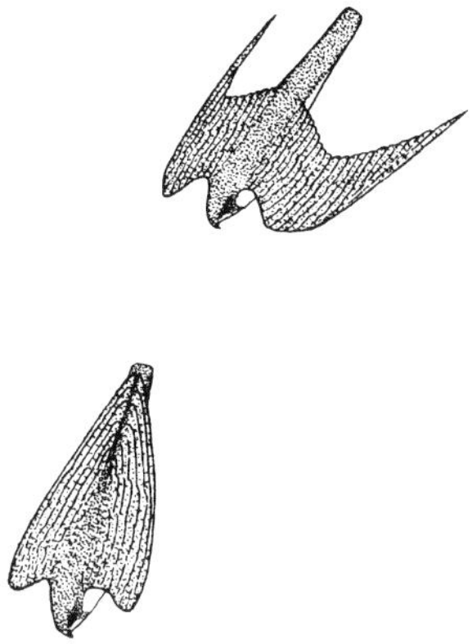

132
133
134

35 136

The Peregrine Falcon (135) has registered a catastrophic decline in numbers throughout the whole world, especially in Europe and North America, over the past thirty years. The shooting of adult birds and disturbing them in their nesting grounds, though these continue to have an adverse effect, are not the main cause — that being the pesticides used in agriculture and forestry. True, these substances effectively exterminate insects and harmful rodents but their poison passes from the insects to the birds that eat them and thence into the body of the falcon causing infertility of the eggs and other adverse effects in the breeding period. As a result there are not enough young birds to counterbalance the mortality rate of the species.
Falco peregrinus

The Saker Falcon (136) breeds from southeastern Europe eastwards to central China. As yet it is not as endangered a species as the Peregrine Falcon. It is very skilful in hunting mammals and birds.
Falco cherrug

Chapter 11 FOWLS AND GAMEBIRDS

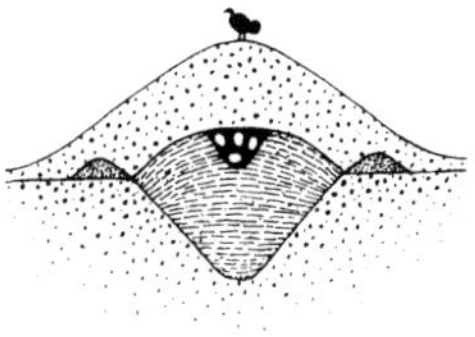

Nesting mound – Brush Turkey's incubator

Many of the most typical characteristics of fowl-like birds are well known for man has shown interest in most of the species of this order since days of old. Besides marked sexual dimorphism, polygamy and the hen's proverbial care of the chicks they are characterized by the fact that they forage for food by scratching the ground. The strong legs are well adapted for this purpose with blunt claws and a short, stout bill, with which larger birds can kill even small animals such as frogs, snakes or mice. However, every rule has its exceptions, particularly in biology, and this applies to fowl-like birds in all respects. The cock is not conspicuously different from the hen in size and coloration in every species, as witness the common partridge or quail; some species live in pairs, e.g. the partridge, guineafowl, curassow. In some species the cock, too, cares for the chicks, in some, on the other hand, not even the hen shows any particular interest in rearing and the young must fend for themselves from the very first (e.g. certain megapodes). And as for scratching the ground for their food, this characteristic applies to all species that feed primarily on the ground but is not developed in those species that live mainly in trees and feed on fruits. Best known of the latter group are the American curassows, guans and hoatzin.

The Black Curassow *Crax alector* (137), like the other thirty-seven species of the American family Cracidae, differs from true gallinaceous birds mainly in that it lives mostly in trees where it also builds its nest. It lays only two to three eggs. The birds live in pairs and after the breeding season they form flocks. In its native

137

138

land, the interior of South America, it is often kept as a domestic bird.

The Razor-billed Curassow (138) is distinguished by an arched beak with comb-like protuberance on the upper bill, an ornament such as is found in no other bird. The sexes have similar plumage. It inhabits the tropical fore ests bordering the Amazon and Orinoco rivers.
Mitu mitu

The Brush Turkey (139) is a member of an unusual family of gallinaceous birds of Australia and the Pacific area which are the only ones that do not sit on their eggs but incubate them in what might be termed hatcheries consisting of large mounds of earth and vegetation, measuring several metres across at the base and more than one metre in height, that often take several months to build. In these the birds excavate holes in which the eggs are laid,
Alectura lathami

139

140

being incubated by the heat generated by the rotting vegetation. The eggs are very large and are deposited by the female at intervals of several days. The male keeps a constant check on the temperature inside the mound, testing it with his tongue, regulating it either by scratching the vegetation away to allow air to penetrate or else adding more rotting material to increase the temperature. The young hatch after six to nine weeks and are developed to such a degree that they are immediately capable of flight. They usually look after themselves from the time they hatch.

The Willow Grouse or **Willow Ptarmigan** (140), of the family Tetraonidae, is a bird of the north. It is of circumpolar distribution in the forest-tundra belt. The winter plumage is white, except in the subspecies found in Britain. It is monogamous. In the autumn, flocks of these birds migrate several hundred kilometres farther south to their wintering grounds, outside the

Lagopus lagopus

141

142

region of the arctic night. They forage for food by digging tunnels in the snow.

The Ptarmigan or **Rock Ptarmigan** (141) ranges even farther north than the Willow Grouse,
Lagopus mutus
nesting even on arctic islands. The isolated populations of the snowfields in the high mountains of the south, e.g. the Alps, Pyrenees, central Asia and Japan, are remnants from the time of the Ice Age.

The Hazel Hen (142) inhabits the boreal zone from Scandinavia to the Sea of Okhotsk and
Tetrastes bonasia
Sakhalin; in the more southerly parts of Europe it is found only in mountain areas. It is fond of mixed woods, preferably with birch and beech trees and an undergrowth of blueberry and cranberry bushes. Its nest is practically impossible to locate for the plumage of the sitting hen blends perfectly with its surroundings; she sits so tenaciously on the eggs that she will not even budge when stroked. The eight to twelve eggs are yellowish with rufous spots. The chicks are capable of flight within a few days of hatching.

143

The Blue Grouse (143) is a North American species the size of the Black Grouse. It is a bird of the woods and stone fields of the Rocky Mountain region. The cock differs from the hen only by its slightly larger size and pronounced orange stripe over the eye.
Dendragapus obscurus

The Black Grouse (144) has a wide range of distribution from Britain, Scandinavia and central Europe across all of central Asia. It is found on the margins of woods, in rushy pastures, on heaths and moors, up to the tree line in high mountains. The cocks' courtship display is an amazing sight. It takes place on the ground in special areas with several cocks performing at one time. The display has two stages and is accompanied by a musical bubbling song. It includes excited hopping about with drooping wings, the tail outspread and cocked upwards, and engaging in fierce combat, mostly symbolic, with the hens looking on. The seven to ten ochre eggs with rufous-brown spots are laid in a nest concealed in heather or blueberry bushes.
Lyrurus tetrix

The Capercaillie (145) is the largest of the wood grouse, found in the woodlands of central and northern Europe as far as the Baikal region. During the past decades this shy bird has completely disappeared from many parts of Europe due to the decreasing number of quiet and undisturbed woodlands. The courtship display, which the impatient cocks begin in the early spring before dawn, is an unforgettable sight. It involves a series of knocking sounds terminated by a 'pop' followed by a brief period when the cock is mute. The hen lays only five to eight fairly small eggs (resembling those of the Black Grouse) on which she sits very firmly.
Tetrao urogallus

144 1

187
188

189

The Red-wattled Lapwing (189) breeds from the southern part of the Arabian peninsula and Mesopotamia across the whole of southern Asia as far as Malaysia. Its behaviour and nesting habits are very like those of the Lapwing, except that it always keeps close to water. It has an ornamental naked yellow wattle at the base of the bill.
Vanellus indicus

The White-tailed Plover (190), a bird with exceedingly long legs related to the Lapwing, is widely distributed in Syria, Mesopotamia and Iran, its range extending north as far as the Caspian and Aral Seas. It breeds in the semi-steppe environment of salt and freshwater lakes. The clutch comprises four pear-shaped eggs the colour of light soil with fine markings. Birds inhabiting more northerly localities migrate as far as East Africa and India.
Vanellus leucurus

The Wandering Tattler (191) inhabits eastern Siberia as far as Kamchatka. Its legs are comparatively short in comparison with those of other waders. The chest and flanks are vermiculated. It nests alongside mountain streams up to elevations of about 1,800 metres and flies great distances to its wintering grounds in Polynesia and Australia.
Tringa brevipes

190

191

The Greenshank (192) *Tringa nebularia*

measuring about thirty centimetres, is one of the largest of this group of birds. It has green legs, a slightly upturned beak and striking white patch extending from the tail to the middle of the back. It breeds from Scotland and northern Europe across the whole of northern Asia as far as Kamchatka. Most birds of this group can be identified by their voice. The Greenshank's is a loud, flute-like whistle that sounds like 'tiuck-iuk-iu'. The birds start out on their migratory flights shortly after they have reared their broods so that northern birds fly through central Europe as early as July. Their final destination is usually Africa; more easterly populations migrate to southern Asia and Australia.

192

193

The Curlew or **Whaup** (193) belongs to a genus numbering eight species of the largest (up to sixty-two centimetres) long-legged waders with brown-spotted plumage and strikingly long, downcurved beaks. The Curlew is a typical inhabitant of moorland and wet meadows. It breeds in the northerly coastal areas of Europe and in a broad belt across central Asia; it is also found in several isolated localities in central Europe. Of all the waders it has the most melodious voice – slow, flute-like whistles and trills which may often be heard not only in spring but at other times as well. It is interesting to note that on hatching the chicks have short straight bills which become longer as they grow.

Numenius arquata

The Black-tailed Godwit (194) is an attractive, slender bird which nests in large meadows near water with shallow muddy shores. In Europe it is found mostly in the

Limosa limosa

194

Black-tailed Godwits copulating

195

areas bordering the North and Baltic Seas and in the more southerly parts of the former USSR as far as the Altai; in central Europe it nests only in larger pond and lake basins. The adults protect the young by flying about the intruder with loud cries, and even feign injury to lure him away.

The Dunlin (195) is the most numerous of the sandpipers, which number some twenty species.
Calidris alpina

The breeding grounds of all sandpipers are located in the northernmost parts of Europe, Asia and America; only the Dunlin's extends farther south to the British Isles and the coast of the North and Baltic Seas. All sandpipers are more brightly coloured in spring, whereas after the summer moult they are a sober grey-brown. During the breeding season the Dunlin has a characteristic large black patch on the belly.

196

197

The Sanderling (196, 197), which has a circumpolar distribution on the northernmost islands and coasts of the Arctic Ocean, is the most lightly coloured of all sandpipers in the autumn. The upper parts are pale grey and the belly white. Groups of these sandpipers run with great speed in typical pattering fashion in the wake of receding waves, quickly picking up all sorts of small animals from the sand with their bills. When migrating the Sanderling, like many other species of waders, stops regularly on the way at certain specific places which are often the scene of vast congregations of birds (197).
Calidris alba

The Ruff (198) is noted for and distinguished by the brightly coloured ruff and ear tufts that adorn the males in spring and by the great fights they wage. The ruff shows such marked variations in colour that no two males are completely alike. After the summer moult there are no differences between the sexes except that the males are bigger. The Ruff is the only sandpiper with longer legs and longer neck which make it look like a redshank. It nests on the southern coasts of the North and Baltic Seas, also in Scandinavia and in the east throughout all of northern Asia. In autumn and spring it migrates regularly across the continent (birds from as far away as the Yenisei River region in Siberia fly across central and western Europe) to its wintering grounds in the Mediterranean region, though more often to tropical and southern Africa.
Philomachus pugnax

The Great Snipe (199), like all eighteen species of snipes and woodcocks, has short legs and a long bill the tip of which is equipped with nerve-endings and can be
Gallinago media

Ruffs engaged in a 'fight'

198

opened while the rest is kept closed thus enabling the bird to locate and grip worms and insect larvae deep in soft soil. The eyes of snipes are located farther back on the head, high up near the crown, so that they can keep on the lookout for danger while foraging for food. The Great Snipe is noted for its interesting mass courtship display in spring which takes place at dusk in the absence of females, at regular courting grounds. During the performance the males clap their bills rapidly and make a series of bubbling sounds reminiscent of a frog chorus. The Great Snipe is found in Scandinavia and the western part of Russia as far as central Siberia. It flies an eastern route to its wintering grounds in East and South Africa; it is rarely seen in central and western Europe.

199

200

201

202

The Common Snipe (200, 201) is a common bird throughout the northern parts of Eurasia and North America. It breeds in wet meadows, marshes and even in moors high in the mountains. In spring the males perform their courtship displays on ground or in the air, their breathtaking flights accompanied by the rhythmic note 'tika tika tika'. Sometimes one may hear a peculiar drumming sound caused by the vibration of the outspread tail feathers. The four eggs, in the well-concealed nest in the grass, are incubated by the female herself, but the male assists in caring for the young. The Common Snipe sometimes has two broods a year. It winters in western Europe bordering the Atlantic, the whole Mediterranean region and Africa as far as the equatorial belt.

Gallinago gallinago

The Woodcock (202) is a popular gamebird traditionally hunted in spring during the courting period. At this time the males fly low above the woodland clearings at dusk sounding a soft 'pseep' followed by a gobbling call 'kvorr-kvorr-kvorr'. Both males and females pair briefly with various partners. The clutch comprises four eggs which are incubated solely by the female, who also rears the young unassisted. If the nest is approached she will sit absolutely motionless. The woodcock is found in damp mixed woods even high up in the mountains throughout all of Europe except the southern and northern parts and in the temperate zone of Asia as far as Japan.

Scolopax rusticola

203

The Avocet (203), like all other members of the family of stilts and avocets, is easily recognized
Recurvirostra avosetta by its long, thin legs, long beak, which may be straight or curved, and the black and white plumage. It inhabits mud flats where it also nests, usually several pairs together. Besides its continuous range in south-eastern Europe and central Asia as far as the region beyond Lake Baikal, it also nests locally in central, western and southern Europe as well as in North and South Africa.

The Red-necked Phalarope (204) is one of three species that form the Phalarope family which
Phalaropus lobatus resembles the sandpipers. They have lobed feet, for they spend most of their time on water. Unlike most birds, the females are bigger and more brightly coloured than the males and incubation and care of the young are solely the responsibility of the males. They breed in the arctic and subarctic zones of the world, wintering on the western coasts of South America, in the Indian Ocean and in the Pacific.

The Pratincole (205) is one member of the family comprising sixteen species of birds mostly of
Glareola pratincola the open plains with sand-coloured plumage, which are found in the warm regions of southern Europe, Asia, Africa and Australia. They are

204

205

excellent fliers as well as runners. They are gregarious birds and also nest colonially. The three eggs are laid on the ground. The insects on which the birds feed are caught mostly on the wing.

The Egyptian Plover (206), a solitary species of the family of coursers and pratincoles, lives along the sandy river banks of tropical Africa. It is noted for its habit of burying not only its eggs but even small chicks in the sand, particularly when danger threatens. It is often seen near resting crocodiles, its cries warning them in case of danger.
Pluvianus aegyptius

The Stone Curlew (207) is a peculiar bird resembling a small bustard, with large yellow eyes that indicate it is nocturnal in habit. It occurs in fairly small numbers
Burhinus oedicnemus

206

207

in the warm, semi-steppe regions of Europe, the Middle East and North Africa. It nests on the dry, stony banks of rivers or in broad sandy and muddy areas covered with low vegetation. It lives in pairs and leads a very inconspicuous life, for it does not become active until dusk. The clutch comprises only two eggs coloured grey with brown spots, which both partners take turns incubating. The diet consists of insects, snails and worms. There are altogether nine species of this family.

The Yellow-billed Sheathbill (208) is an Antarctic wader the size of a dove, with plump body, snow-white plumage and saddle-like sheath on the bill that covers the nostrils. It is equipped for survival in its harsh environment with a thick coat of down and thick layer of subcutaneous fat. It feeds on small animals, carrion and vegetable food cast up on the seashore. There are only two species of sheathbills.

Chionis alba

208

Chapter 14 THE RAUCOUS TRIBE OF GULLS AND TERNS

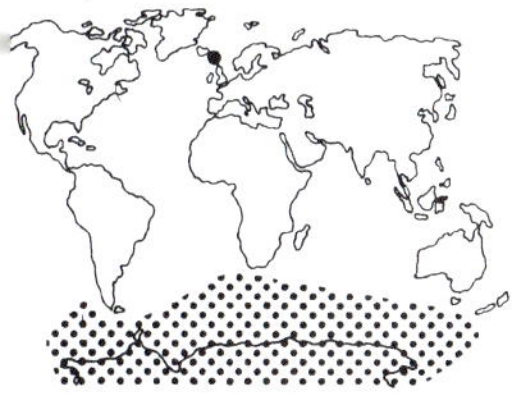

Range of the Great Skua

The neighbourhood of water, be it the shores of lakes, ponds, marshes or oceans, always has a characteristic background noise of bird calls and cries. The fortissimo part of this chorus is almost always performed by gulls and terns. Every bird has an extensive repertoire of shrill calls and cries, their effect magnified by the fact that these birds usually live in large colonies, particularly during the breeding period.

Gulls, terns, skuas and skimmers belong to the same order (Charadriiformes) as waders and alcids, with whom they have many anatomical features in common. Nevertheless, of the order's 334 species the 91 species of gulls, terns, skuas and skimmers form a uniform group (Lari) that is easily distinguished from the rest. Even the layman knows what a gull looks like and even if he may not recognize a tern as such he is certain to think it is a gull and, from the systematic viewpoint, he will not be far from the truth. Members of this group are medium-large to large birds. They do not weigh much — two kilograms at the most — but they are impressive in flight for their prolonged wing quills give them a greater wingspan. They moult twice a year, the breeding plumage differing from the winter plumage. That they are birds bound to water is borne out by the webbed feet; however, the courtship antics, which exhibit great differences among the individual species, are performed mostly on land. These birds have always attracted attention with their widely diverse behaviour, both during and out of the breeding season. Some forms of expression are common to all, but in addition to this each species has many more of its own innate forms. They breed once a year, building their nests on the ground near water. The young scamper about shortly after hatching but remain on or near the nest for a long time, being fed by the adult birds from beak to beak. Animal food is the mainstay of the diet.

The Great Skua (209) *Stercorarius skua* is a member of the family Stercorariidae, birds that differ from gulls and terns in having darker, brown plumage, at least on the upperside. It is

209

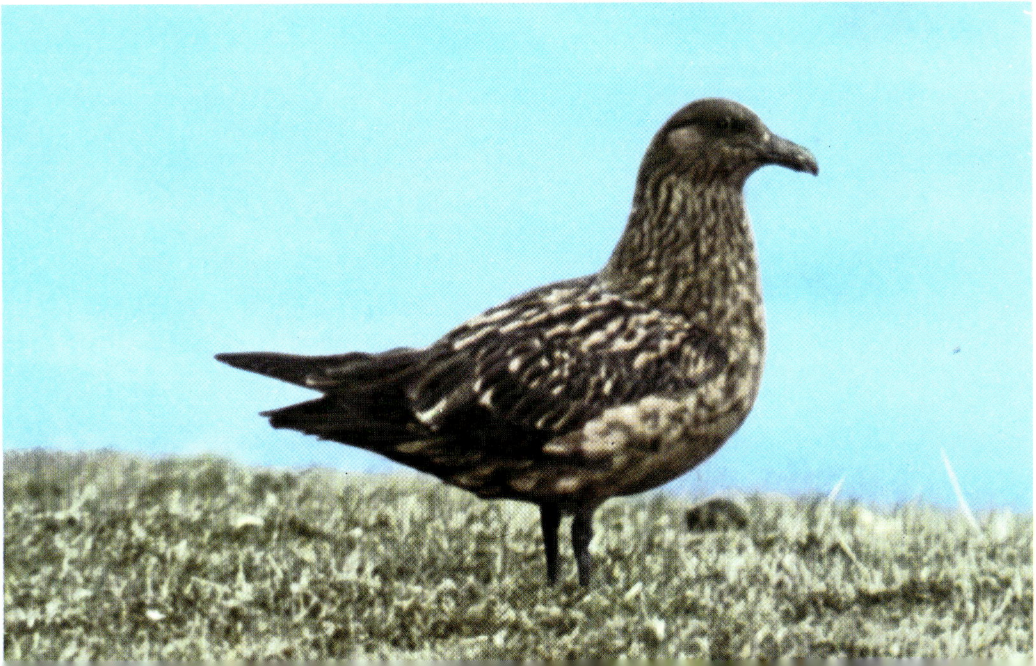

211

210

distinguished by a strong, hooked bill composed of four horny sections. It is an expert flier and takes advantage of this fact to obtain food from other birds by swooping down upon them until they drop their catch, which it retrieves adroitly before it falls to the ground. During the breeding season it robs the nests of other birds, taking both eggs and young. It generally lays two eggs in an untidy scrape in the ground, with both parents taking turns incubating for three to four weeks. The Great Skua is the largest member of the family. Of interest is its distribution which comprises two distinct areas — one in the Antarctic and the other in the northern hemisphere in Iceland, the Faroes, Shetland and Orkney islands and in northern Scotland. In the Antarctic it fulfils the role of raptorial birds, which are not found here.

The Arctic Skua (210) is much smaller than the Great Skua. It inhabits the islands of the northern seas and the tundras of Eurasia and America. It occurs in two colour phases — all brown, found mostly in the southern parts of its range, and brown with pale underparts, found mostly in the north. The central tail feathers are pointed and project about nine centimetres beyond the rest of the tail.

Stercorarius parasiticus

The Long-tailed Skua (211) is easily identified in flight by its long, narrow central tail feathers which project sixteen to twenty-five centimetres beyond the rest of the tail, and by its smaller size. It has a discontinuous distribution in the lowland and mountain tundras of Eurasia and America. Singly or in small groups these birds roam the countryside in graceful flight seeking their prey close above the ground. The Long-tailed Skua is not as much of a predator as the other skuas, feeding mostly on lemmings, insects and various small animals. The two young nestlings, which abandon the nest two days after hatching, are fed mainly insects.

Stercorarius longicaudus

The world is inhabited by a total of forty-three species of gulls. The various species differ more in size than in coloration. White and grey predominate; most gulls have a grey or black mantle, others a black head or at least a black mask. It is very difficult to distinguish between various species of the same size according to the plumage of the immatures for they are coloured a sober brownish-grey. Gulls acquire their adult plumage after two to four years, depending on the size of the species.

The Red-legged Kittiwake (212), a very attractive and dainty gull, breeds on certain of the islands in the Bering Sea. It nests either in colonies of its own kind or in the company of other marine birds. It looks very much like the Black-legged Kittiwake but the mantle is darker grey, the legs are red and the bill is comparatively short, as its scientific name indicates. Its nesting habits are much the same as those of the Black-legged Kittiwake. The nest is built of dry grasses, moss and seaweeds with an admixture of mud. The clutch generally consists of two eggs, which both parents take turns incubating for twenty-four to twenty-six days.

Rissa brevirostris

212

213

The Black-legged Kittiwake (213) has black legs and a grey mantle. Both members of the genus *Rissa tridactyla* — *Rissa* have no hind toe. It is very plentiful and often the most numerous of the inhabitants of the coastal cliffs and islands of the north Atlantic and Pacific oceans. The nests are usually packed tightly together on narrow cliff ledges. Associated with this method of nesting are a number of specific characteristics which distinguish the Black-legged Kittiwake from other gulls that nest on flat or sloping ground. The young remain immobile in the deep hollow of the nest, do not scatter, and do not beg for food by 'pumping' their heads up and down but keep them turned towards the cliff face. Outside the breeding season they spend their time on the open sea.

The Black-headed Gull (214) has a chocolate-brown head in the breeding season; in winter the *Larus ridibundus* — head is white with a brown spot in the region of the ear. The Black-

214

headed Gull nests in large colonies and is primarily an inhabitant of inland bodies of water throughout the temperate zone of Eurasia.

The Glaucous-winged Gull (215) is one of the large species of seagulls. It inhabits the islands and coasts of the North Pacific. Except for its grey wing tips, it resembles the Herring Gull in size and coloration.

Larus glaucescens

215

216

The Herring Gull (216, 218) inhabits the coasts and some inland lakes throughout most of the northern hemisphere. Its silhouette – the wingspan measures 1.5 metres – may be seen in the wake of ships far out to sea. It nests usually in colonies, sometimes numbering thousands of pairs, on flat grassy islets and steep coastal cliffs, amidst the tough vegetation of coastal sand dunes, in reeds, stone debris and, in some coastal towns in the Balkans also on the roofs of houses. The female lays two or three eggs coloured olive brown or green with dark spots. The parents take turns incubating for four weeks. The young, covered with speckled down on hatching, soon abandon the nest but remain in its vicinity for four weeks. It was discovered that the adults' yellow bill with red spots plays an important role in caring for the young. Together with the sound of the parents' voices it provokes begging movements in the young which in turn rouses the feeding instinct in the adults. The courtship performance is very striking and the bird has a very wide range of calls. It feeds on all marine animals it is able to handle and also eats carrion as well as the eggs and young of other birds. In the case of overpopulation it causes great damage.

Larus argentatus

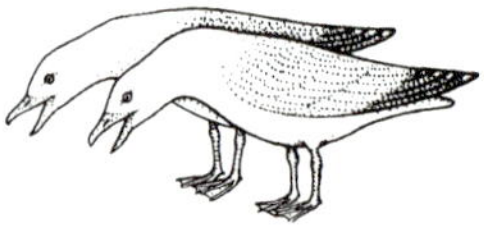

One stage of the Herring Gull's courtship performance

The Mediterranean Gull (217) resembles the Black-headed Gull. It is only slightly larger and in the breeding season has a jet-black head and white wing tips. The nest is located in thick coastal vegetation. The Mediterranean Gull breeds in southeastern Europe though it also nests irregularly in central and northern Europe and isolated nesting grounds are also to be found in the Gobi Desert. It forages for food – gastropods, insect larvae, worms – in flocks which fly to fields and steppes far from the nest.

Larus melanocephalus

217

218

219

The Ross's Gull (219) is a small gull distinguished from the others by its wedge-shaped tail, rosy
Rhodestethia rosea plumage and narrow, black neck-band. It nests in northeastern Siberia.

The Great Black-backed Gull (220), largest of all the gulls, measures up to 76 cm and has
Larus marinus a wingspan of 170 cm. It nests on North Atlantic coast, usually in the colonies of other marine birds which often fall prey to it. It steals their eggs and young and even kills adult birds with its strong beak.

Terns differ from gulls mainly in that they are more slender. They have very long, narrow wings, a slender, pointed bill, graceful legs which are short in proportion to their size and a comparatively long, forked tail. Unlike the gulls, most of the forty known species of terns inhabit the warm and temperate zones. Even the few species that nest in the north spend the greater part of their life in warm winter quarters. Though their way of life is attached to water they rarely alight on the surface. They feed mostly on fish, plummeting into the water after them in much the same way as gannets and tropicbirds.

220

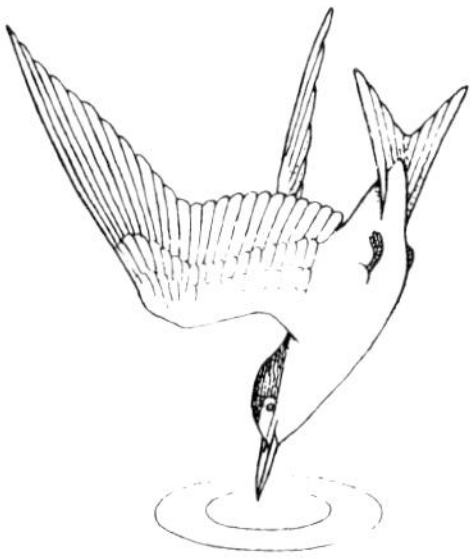

The Little or Least Tern *(Sterna albifrons)* **plunging after a fish**

221

The Arctic Tern (221) has a range of distribution that extends far to the north. It breeds in North America and Eurasia on sea coasts and islands, but also even far inland in the tundra. Birds nesting in North America cross the Atlantic and fly along the coasts of Europe and Africa as far as the Antarctic seas. Some even fly from the west coast of Africa to the shores of South America and farther south. The trip there and back covers thirty-five to forty thousand kilometres – a truly remarkable feat.
Sterna paradisaea

The Caspian Tern (222), largest of all the terns, has a remarkably discontinuous distribution. Its nesting colonies are located in several isolated areas in all parts of the world – in North America, Africa, southern and northern Europe, central, southern and eastern Asia, Australia and New Zealand. The clutch comprises two to three eggs. The picture shows one of the stages of its courtship display.
Hydroprogne caspia

222

223

224

The Common Tern (223) greatly resembles the Arctic Tern but has a coral-red bill tipped with black and a shorter tail. It inhabits Europe, Asia and North America and undertakes long flights to its winter quarters in southern Africa and America. It nests in open places along seashores as well as on the shores of inland lakes, ponds and rivers. It spends most of the daylight hours in the air and is clumsy on land. Colonies of terns are extremely noisy. The courtship display of the Common Tern takes place both on the ground and in the air, when the male, carrying a small fish in his bill, is followed high in the air by the female to the sound of raucous cries on the part of both. The eggs are coloured brownish-green with grey and

Sterna hirundo

225

dark spots, and are incubated by both parents for twenty to twenty-two days. The two to four, newly-hatched chicks remain with heads resting on the edge of the nest for a few hours but soon after abandon it and conceal themselves in the surrounding vegetation, returning now and then for about two more days before taking permanent leave. The young are very good swimmers.

The White-fronted Tern (224) has a striking white forehead. It breeds on the eastern coast of Australia, in Tasmania, New Zealand and certain other islands. The breeding season in these latitudes is in November and December.
Sterna striata

One often comes across evidence that the shape of the bill corresponds to the method whereby a bird procures its food. This is demonstrated also by the bill of the skimmers, which are close relatives of the terns. It is long, straight and laterally compressed, with the lower mandible far longer than the upper. The upper bill has a sharp cutting edge which fits tightly into a shallow groove in the lower bill. At dusk skimmers fly close to the water with the bill open, the lower mandible skimming the surface, gathering small fish and other animal food. Skimmers have another peculiar characteristic – their pupils are vertical, slit-like. They nest on the flat shores of the sea coast as well as inland waters.

The Black Skimmer (225) with red and yellow bill tipped with black makes its home in America.
Rynchops nigra

The African Skimmer (226) with yellow bill, is found on the waters of tropical Africa.
Rynchops flavirostris

226

227

The Whiskered Tern (227) is mostly grey with black cap and white cheeks. It nests in smaller colonies in reed beds and feeds mainly on fish. It has an interesting, discontinuous distribution – occurring in southern and eastern Europe, Africa, southwestern, southern and eastern Asia and also in Australia.
Chlidonias hybrida

The Inca Tern (228), found on the coasts of Peru and Chile, is a bluish silvery-grey bird with yellow, leathery flaps of skin under the eyes, red bill and feet. It nests in cavities – under rocks or in holes in the ground. It sometimes picks meat from between the teeth of sea lions.
Larosterna inca

228

Chapter 15

CLIFF-BREEDERS OF THE NORTHERN SEAS

Birds that hunt their food only on the open seas, where they spend both day and night, must become temporary land residents at least during the breeding period. In the northern seas, in places with an abundance of food, some species of marine birds congregate in large nesting colonies on cliffs — a curiosity of the bird realm in the northern hemisphere since days of old. The twenty-three species of alcids (family Alcidae), which exhibit the greatest dependence of all birds on the sea, are sometimes joined by a mixed company of gulls, cormorants, gannets, shearwaters and petrels, etc. The alcids are closely related to waders and gulls and in appearance greatly resemble the penguins of the southern seas. Unlike the latter, however, they are good fliers. They are expert divers and fine swimmers, propelling themselves on the surface with their short, webbed feet and underwater with the aid of their wings, which, however, are not as efficient oars as those of the penguins; underwater the feet function as a rudder. Some alcids remain underwater as long as two minutes, exceptionally even longer, and dive to depths of thirty metres. The sea is the source of their food supply — fish, crustaceans, worms and other invertebrates. Most species lay only a single egg either directly on the ground or rock or else in rock crevices, under stones, or in burrows which they dig with their beaks and feet. Both parents incubate and care for the young. The young often leave the nest, even though they have not yet acquired their flight feathers, and flutter down from the high cliffs either to the water or the shore. Few are killed in the process for their outspread wings break their fall and their thick feathers soften the impact

229

230

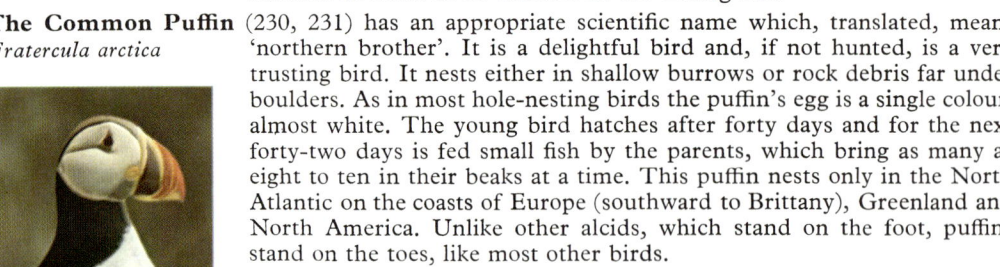

of landing. As soon as they attain full maturity in two to four years' time they return to the same breeding sites from which they started out.

The Tufted Puffin (229) has a striking beak, laterally compressed and gaudily coloured during the breeding season. It nests on Asian and American coasts of the North Pacific, southward as far as California. As many as 100,000 pairs were counted in some of its colonies in the Bering Sea.

Lunda cirrhata

The Common Puffin (230, 231) has an appropriate scientific name which, translated, means 'northern brother'. It is a delightful bird and, if not hunted, is a very trusting bird. It nests either in shallow burrows or rock debris far under boulders. As in most hole-nesting birds the puffin's egg is a single colour, almost white. The young bird hatches after forty days and for the next forty-two days is fed small fish by the parents, which bring as many as eight to ten in their beaks at a time. This puffin nests only in the North Atlantic on the coasts of Europe (southward to Brittany), Greenland and North America. Unlike other alcids, which stand on the foot, puffins stand on the toes, like most other birds.

Fratercula arctica

231

Joint colony of Common Guillemots and Kittiwakes (232)

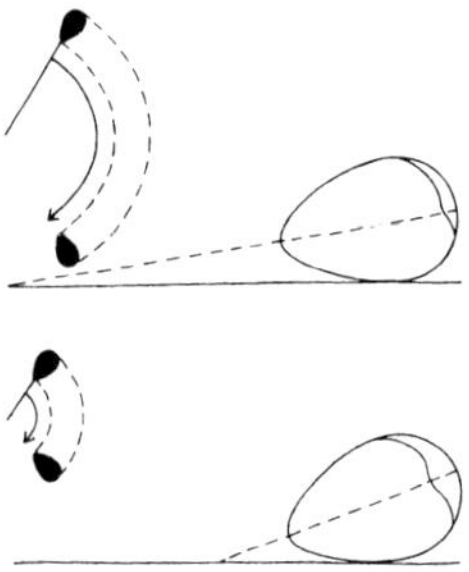

Rotation of a freshly laid alcid's egg and of one that is near the end of incubation

233

The Common Guillemot (232, 233) nests in the North Atlantic (southward to Brittany) and North Pacific. The single egg, showing marked variation in colour, is laid in May to July on a bare rocky ledge. It is cone-shaped so that it does not roll off the ledge but rotates in a small circle round the pointed end,

Uria aalge

234

235

a characteristic which, together with the unusually thick shell, is an effective adaptation to the narrow rocky ledges on which it is deposited. In late April guillemots return from the sea to their breeding sites where they perform their courtship antics jointly both on the shore and cliffs of the islands as well as on the sea. On the water they congregate in large flocks which fly in circles round each other and dive together. The nesting grounds, at this time, are the scene of many battles as the birds fight for choice sites. At about the age of twenty-five days the young leave the nest and set out to sea where they remain in the company of the parents for some time.

The Black Guillemot (234) sometimes establishes its nesting colonies some distance inland but always within reach of the sea. It nests in the North Atlantic and Pacific. The female lays two eggs in cliff crevices or under rocks.
Cepphus grylle

The Razorbill (235) – a smaller relative of the flightless and extinct Great Auk *(Pinguinus impennis)*, nests on the coasts of the North Atlantic and adjoining arctic seas and is the only one of the alcids to do so occasionally even on brackish and fresh waters. It has even been known to nest on an inland lake (Ladoga). The single egg is incubated by both parents for thirty-five to thirty-six days. At the age of about three weeks the as yet immature young bird deserts the nest and flutters down to the water where it is in its element. The mainstay of the diet are small fish, in search of which the Razorbill flies as far as twenty kilometres from its nesting site. The greatest number of alcid species are to be found in the North Pacific. The following are two of the lesser known forms.
Alca torda

The Crested Auklet (236) of the coasts of Chukotski Peninsula, the Kuril and Aleutian Islands and Sakhalin. It feeds mostly on crustaceans and lays its single egg in the month of June.
Aethia cristatella

The Whiskered Auklet (237) makes its home in the southern parts of the Bering Sea – on the Kuril, Komandorski and Aleutian islands as well as on Bering Island. These two species are not found in great abundance on the breeding cliffs of the northern seas.
Aethia pygmaea

236

237

Chapter 16 HAPHAZARD NEST-BUILDERS

Adult pigeon feeding its young

Probably not a single one of the 230 species of pigeons (order Columbiformes) builds a sturdy and attractive nest. Pigeons are definitely not skilled builders. Some nest in hollows and their one to two eggs, as in most hole-nesting birds, are white. However, because even those pigeons that build simple nests of twigs in trees, on the ground or on cliffs lay pale coloured eggs without markings, it is believed that the precursors of today's pigeons originally nested in hollows. The feathers of all pigeons are loosely attached in the skin and easily torn out. The individual feathers are downy at the base, the fine powder produced there serving to cover and protect the feathers in place of the oil of the atrophied or absent uropygial gland. Pigeons are first and foremost tree-dwellers, though some species spend most of the time on the ground. They feed on fruits, seeds and young shoots. They are nidicolous birds, i.e. their young are blind on hatching and are fed by the parents from the crop, the nestling poking their soft beaks inside the throats of the adult birds. The first food is so-called 'pigeon's milk' — a cheesy curd which is the sloughed-off lining of the crop; later the young are fed softened seeds. Pigeons drink by sucking up water; they do not raise their heads to let it run down into the throat as do other birds.

The Crowned Pigeon *Goura cristata* (238) and its related species of New Guinea are the largest of all living pigeons. It spends most of the day on the ground under the canopy of trees, but at night it roosts in trees where it also nests.

238

239

The Nicobar Pigeon (239) inhabits the forests of the East Pacific islands. It is a good flier. It collects food on the ground and nests in trees. The long, narrow neck hackles form a sort of mane.
Caloenas nicobarica

The Thick-billed Green Pigeon (240) is one of many species of so-called fruit pigeons, widely distributed throughout the forests of the tropics and subtropics. Green is the predominant colour common to all. The Thick-billed Green Pigeon is found in southern Asia, Malaya and the Sunda Islands.
Treron curvirostra

240

The Pied Imperial Pigeon (241) is another of the known species of fruit pigeons. Like the others it seldom comes to the ground, feeding, as it does, on the fruits of trees. Smaller fruits are swallowed whole and the bird is able to open its beak wide to stuff its large crop full. Flocks of these contrastingly coloured pigeons inhabit the forests of the islands stretching from the Bay of Bengal to New Guinea.

Ducula bicolor

The Collared Turtle Dove (242), originally an inhabitant of southern Asia and China, later Asia Minor and the Balkan Peninsula, is known primarily for its rapid spectacular spread throughout most of Europe during the past thirty years where it has become a common bird of gardens, parks and tree avenues in the vicinity of human habitations. In 1912 it was recorded for the first time in Serbia, the 1930s marked the start of its spread to Hungary, in 1938 it crossed the border into Czechoslovakia, in 1943 it showed up in Vienna, in 1946 it was already nesting in Germany, in 1949 in Holland, in 1951 in Sweden, in 1955 in England, in 1957 in Estonia and in 1964 it appeared in Iceland. The course of this expansion to the west and northwest was closely observed by many ornithologists thus enabling the drawing of a detailed map which shows, among other things, that large rivers were the main routes of expansion. The Collared Turtle Dove extended its boundaries some two thousand kilometres to inhabit a new area covering some two million square kilometres. The rapid settling of new areas was made possible also by the Collared Turtle Dove's fecundity, for if conditions are favourable it has as many as five broods a year, sometimes nesting even in the winter months. It is not particular in its choice of nesting site, building its haphazard nest of twigs and grass, often even of wires if nothing else is available, in treetops and bushes, on telegraph poles, window sills, etc. In its breeding territory it continually sounds its monotonous, three-syllable cooing note. In the winter the Collared Turtle Dove often congregates in large flocks, particularly where there is an abundance of food — in the vicinity of granaries, mills, farms and poultry farms.

Streptopelia decaocto

Stages of the Collared Turtle Dove's spread throughout Europe

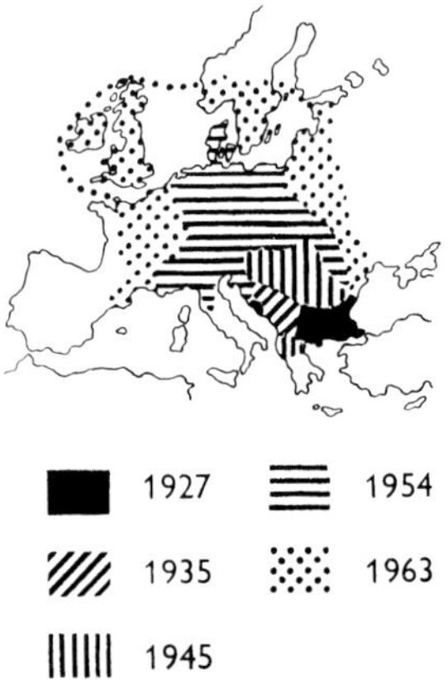

241

242

243

The Galapagos Dove (243) breeds only in the Galapagos Islands, building its nest in trees as well as on cliffs. It is very tame and in the days when it was much more plentiful in its native habitat than it is today it could be killed with sticks and stones, as Charles Darwin wrote.
Zenaida galapagoensis

The Crested Pigeon (244) is an ornamental Australian pigeon often kept as a cage bird. It builds a simple nest in trees and the eggs are incubated eighteen days mostly by the female, the male sometimes relieving her during the day.
Ocyphaps lophotes

244

245

The Turtle Dove (245) is a common inhabitant of open woodlands in Europe, most of Asia and North Africa. Though the nest is built by both partners it is extremely simple — just a thin layer of twigs and grass in the branches of trees or bushes, so thin that often the two eggs inside may be distinguished from below. During the breeding season one may often hear the monotonous cooing of the male perched on a branch near the nest, where he inflates his crop and inclines his head as he coos.
Streptopelia turtur

The Mourning Dove (246) is a common dove of North America and the West Indies. In the southern states of the USA it is hunted as a game bird. It is a migrant; the northernmost populations winter farthest south.
Zenaidura macroura

The Wood Pigeon (247) is a common bird of the forests of Eurasia and North Africa. The nest, a flimsy structure of twigs, is usually located in treetops near the trunk. It nests two to three times a year, always laying two white eggs. In
Columba palumbus

246

247

western Europe it has become a city bird during the past several decades. It often nests in parks and gardens where it strolls around on the grass. In central and northern Europe as well as in Asia, however, it has remained a shy, woodland bird. Wood Pigeons from northerly areas are migratory. In September and October they form large flocks and on clear autumn days fly to the south or southwest, where they join their fellows that breed in these latitudes and are resident. Flocks of migrating wood pigeons, sometimes numbering several thousand birds, alight in stubble fields or beside suitable watering places, sometimes remaining there a number of days. The Wood Pigeon is a popular game bird.

The Speckled Pigeon (248) is one of the common African species of pigeons. It is found in the savannas and on the margins of forests. The breeding season is from November till April and the simple nest of twigs and dried grass is built in trees.
Columba guinea

248

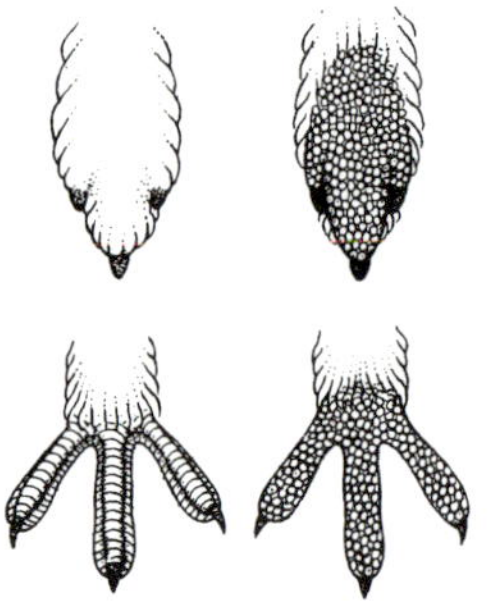

Foot of the Pallas's Sandgrouse *(Syrrhaptes paradoxus)* (top) and that of the Spotted Sandgrouse (bottom)

249

The Spotted Sandgrouse (249) inhabits the desert regions of North Africa, Arabia, south as far as India.
Pterocles senegallus

The Pin-tailed Sandgrouse (250) nests in North Africa, the Iberian Peninsula, the Near East and the Middle East. Sandgrouse are birds of the steppe and desert. Their extremely short legs, and in most cases also their toes are feathered. The toes have horny plates on the underside, apparently an adaptation to running on burning sand. The spotted eggs are laid in an unlined scrape in the open. The young are able to feed themselves soon after hatching. The parents bring them water from distant water holes by soaking the feathers on their bellies. Previously it was believed that sandgrouse suck up water like pigeons but it has been proved that the majority drink water like other birds. The inclusion of sandgrouse in the order of pigeons is today a subject of debate. Their biology and certain anatomical features support the more recent opinions that they are related to waders.
Pterocles alchata

250

Chapter 17 TALKING BIRDS

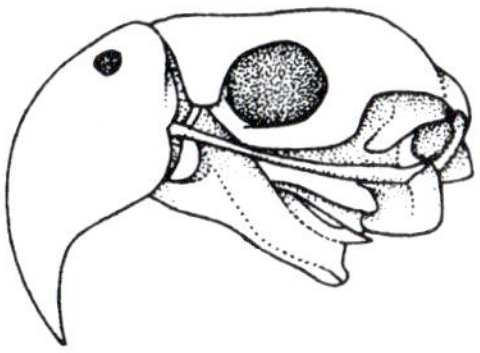

The upper bill of parrots is joined to the skull by a grooved joint which allows for limited movement

The ability to mimic human speech is not limited to a single group of birds, but the first to come to mind are always the parrots. Not all of the 326 species that make up this group, however, are able to reproduce human speech. There is nothing remarkable about the voices of parrots. In most instances they are raucous birds whose shrieks and squawks are definitely not balm to the ear. Parrots are by nature good mimics, easily learning to imitate the voices of other birds and animals as well as various mechanical sounds. However, only certain species raised in captivity are able to reproduce human words. Some species are more talented, others less so, and even within a given species one will find individuals with a greater or lesser ability to learn. In any case, the ability to mimic speech is evidence of high intelligence and good memory. The most gifted in this respect are the Amazon Parrots, Cockatoos and the African Grey Parrot. Certain small Australian budgerigars are quite talented as well. The intelligence and memory of some individuals is so great that they are able to use certain words or phrases on pertinent occasions. These are vocal expressions that have been learned as a response to specific situations.

Parrots and their allies are a uniform group of birds with a down-curved, hooked bill, the upper mandible joined to the skull by a grooved joint, and strong, grasping feet with the first and fourth toe pointing backwards. When climbing they use their beaks to help pull them up; their feet are used to manipulate food and carry it to the beak. Inside the beak is a strong fleshy tongue which aids them in their ability to mimic words. Parrots generally nest in tree holes, in semi-cavities and, in rare instances, on the ground. They inhabit the tropical and subtropical zone in Australia, America, Africa and Indo-Malaysia.

51

252

253

The Kea (251) of New Zealand is distinguished from most parrots by its remarkably long upper bill and by the fact that it lives mainly on the ground. It feeds on plant and animal matter. The males are polygamous.
Nestor notabilis

The Galah or **Roseate Cockatoo** (252) is one of seventeen ornamental, medium-sized cockatoos distributed from the Philippines to Australia and Tasmania. Unlike its crested relatives the Roseate Cockatoo has a fairly short crest composed of broad, flat feathers. It is common in the woods of northern and western Australia and quite unpopular with farmers when large flocks of these birds raid their plantations. The female lays four to five eggs in a tree hole lined with leaves.
Cacatua roseicapilla

The Major Mitchell's or **Pink Cockatoo** (253) has a magnificent fan-shaped crest of scarlet white-edged feathers. In Australia's forests it feeds on plant seeds and with its fairly small bill digs roots and tubers from the ground. The breeding season is from September till December, when it lays three to four eggs. Paired birds return to the same nesting hole for years on end. Major
Cacatua leadbeateri

254

Mitchell's Cockatoo is not very plentiful in its native land and does not form flocks. Like the Roseate Cockatoo it is a popular cage bird.

The White-crested Cockatoo (254) inhabits the northern and central Moluccas and very little is known of its way of life. The strong beak easily cracks the hard shells of fruits it seeks in the treetops and on the ground. Its flight is said to be quite noisy. In the picture we can clearly see that the upper mandible fits inside the broader lower mandible, a characteristic common to all cockatoos.

Cacatua alba

Lories are the most brilliantly coloured group of parrots and therefore popular cage birds. They have a smooth, slender bill that does not have the sharp notch for cracking hard fruits found in most other parrots. The diet consists of soft food and lories are even able to feed on liquids, their tongues being tipped with a brushy fringe with which they lap up the nectar of flowers and the fleshy parts of soft fruits.

256

The Lorikeet (255) is found in large numbers in Celebes and the neighbouring islands and is a popular cage bird with the natives.
Trichoglossus ornatus

The Turquoise Parrot (256) belongs to a group of seven species of small, so-called grass parakeets that inhabit the bush and wooded areas of Australia. All have a fairly small beak and forage for food on the ground. They nest in the hollows of eucalyptus trees high above the ground. The Turquoise Parrot was at one time very plentiful in eastern Australia but is now rare.
Neophema pulchella

The Chattering Lory (257) is likewise a popular cage bird. It, too, feeds mostly on pollen and the nectar of flowers. It is a native of the northern Moluccas.
Domicella garrula

257

255

The Budgerigar (258) is without doubt the most favoured pet of all. The wild form, still found
Melopsittacus undulatus

in vast numbers in the interior of Australia, is coloured yellow-green and grassy-green barred with black. Since 1840, when Budgerigars were first introduced to Europe, breeders have developed many colour varieties, the most common being blue and yellow. In the wild the Budgerigar feeds on the seeds of semi-desert grasses. As a cage bird it is easily cared for and its voice is definitely more pleasant than that of larger species. Some birds become tame and learn quite a number of words and even whole sentences – a delightful thing in such a small bird. It nests in the hollows of trees, often in large colonies.

The Little Vasa (259) of the group of blunt-tailed parrots (Psittacini), is one of the few species
Coracopsis nigra

with sober, brown-black plumage. It is a native of Madagascar, and is rarely found in zoos. It lives in colonies in treetops and is believed to lay only two eggs.

The Monk or **Loro** (260), with its sober green and grey plumage, does not stand out among the
Myiopsitta monachus

other, more gaudily coloured parrots, but it is noteworthy for its nesting habits. In the forests of southern Brazil, northern Argentina and Bolivia, numerous pairs of birds gather during the breeding season to build large communal nests in the tops of trees. These measure as much as three metres in diameter, with each pair of birds having its own compartment with separate entrance. After the breeding season is over the birds form large flocks which sometimes cause great damage to South American plantations.

The Upper Yangtse or **Chinese Parakeet** (261), one of the twelve species of long-tailed para-
Psittacula derbyana

keets found in Africa and Asia, inhabits the warmer areas of southwestern

259

260

261

262

263

China, whence it extends to the pine forests and rhododendron growths of Tibet, up to elevations of more than three thousand metres. It feeds on the seeds of herbs and trees.

The Pennant's Parakeet or **Crimson Rosella** (262) is a typical representative of the seven species of flat-tailed parrots inhabiting the open lands of Australia. In eastern and southeastern Australia it is a very common bird that may be found even in city parks. The female lays five to eight eggs in the hollows of trees.

Platycercus elegans

The small, gaily coloured lovebirds (genus *Agapornis*) are very popular cage birds. Individual pairs of these fairly large-headed parrots are practically inseparable both in the wild and in captivity. When startled or in danger they perch on a branch in a row, one beside the other. They generally nest in cavities but some species build a spherical nest.

The Masked Lovebird (263) carries material with which to line the nest — stems, twigs and pieces of bark — in the bill, whereas other species insert it in the tail, back or neck feathers and shake it loose inside the cavity. Shown in the picture is the form *Agapornis personata fischeri* found in eastern Tanzania.

Agapornis personata

264

The Blue-fronted Amazon Parrot (264), a native of the forests of Brazil and likewise often
Amazona aestiva

kept as a cage bird, is coloured green with a yellow head and blue forehead. Like all other twenty-five species of Amazons it is an excellent climber but is clumsy in flight and on the ground.

The African Grey Parrot (265) is not particularly handsome; it is pale grey with a red tail. It is
Psittacus erithacus

popular, however, because of its remarkable powers of speech. In good hands, and with infinite patience on the part of the keeper, it can be taught dozens of words and sentences. In captivity it commonly reaches the age of fifty years. After the breeding period, which takes place during the rainy season, Grey Parrots congregate in large flocks which are extremely noisy. In the wild, however, they are not known to mimic other creatures. They inhabit the forests of Africa between Guinea, Angola and Lake Victoria.

265

266

267

268

269

The Scarlet Macaw (266) is a member of the genus *Ara*, which are the largest parrots of all. Each, in its way, is beautifully coloured and all have long, wedge-shaped tails, so that some species measure as much as one metre in length. The Scarlet Macaw is a native of tropical South America, its range extending as far as Mexico.
Ara macao

The Military Macaw (267) inhabits wooded areas in the lowlands and mountains of Mexico, Colombia, Bolivia and Peru. It forms small groups, nests in tree holes and lays two eggs.
Ara militaris

The Gold-and-blue Macaw (268) inhabits the woods and virgin forests from northern Argentina to Panama. With its strong beak – the same as other species of macaws – it easily cracks even hard fruits, first filing the shells in one spot with the hard notches of its upper bill. Even hard nuts yield to this instrument. Of all the macaws, this species is the one most commonly kept in captivity.
Ara ararauna

The Hyacinth Macaw (269) differs markedly from the other macaws with its feathered face, cobalt blue colouring and large size (980 millimetres in length). It lives in pairs or small groups in central and eastern Brazil, feeding, like its relatives, on fruits and small animals; the mainstay of the diet, however, is the nuts of palm trees. The Hyacinth Macaw nests in burrows which it excavates with its beak. It lays two to three eggs, the young hatch after twenty-five to twenty-eight days and leave the nesting hollow after about three months. It is highly prized by keepers for its unusual colouring. Its powers of speech, however, are not very good.
Anodorhynchus hyacinthinus

Chapter 18 A SMALL BIRD POPULATION FOUND ONLY IN AFRICA

Range of mousebirds

Africa is the home of a great variety of birds found on no other continent. This applies mainly to the area south of the Sahara where the mousebirds live on the margins of forests, in vegetation bordering rivers as well as in the bush. They are interesting, agile birds which are somewhat larger than songsters. All six species raise their crests upright when upset and all have long, graduated tails that are much longer than their bodies. Grey or brown, like mice, they scurry among the thickets and hop about close to the ground as well as in the tops of bushes. This habit, along with their remarkably fine, hair-like feathers, is what gave them their name. Mousebirds exhibit many peculiarities both in anatomy and biology. One of the structural peculiarities is the foot, which has two reversible toes (the first and fourth) that can be used forwards or backwards. When climbing or hanging — at which mousebirds are past-masters — all four toes point forward; when moving along branches two are turned forward and two backward. The young are remarkably small. When newly hatched they weigh about two grams and even after they have fledged and have left their basket-shaped nests they are still about

271

half the size of the parent birds. Because of these and other unusual features they are placed in an order by themselves (Coliiformes). The diet consists of seeds and fruits, occasionally also insects.

The Blue-naped Mousebird (270) is the longest of all the birds of this family – it measures 350 millimetres. It breeds in the bush of central Africa. As a bird of dry, arid country, it is not bound to water. The water contained in the fleshy fruits it feeds on is sufficient for its needs and dust or sand bathing takes care of its feathers.

Colius macrourus

The Bar-breasted Mousebird (271) is widespread from Nigeria to Ethiopia. It inhabits woody areas, savannas and mountains up to 2,000 metres. Like other mousebirds it leads a social life, nesting in small communities.

Colius striatus

Chapter 19 CUCKOOS AND THEIR KIN

Laying eggs in the nests of other species and leaving the chores of incubation and caring for the young to the unfortunate hosts is a well-known habit of the Common Cuckoo *(Cuculus canorus)*. This habit is attributed by man to all cuckoos, but such is not the case. In reality the cuckoo's immediate relatives (members of the order Cuculiformes which embraces some 130 species) show marked differences in this respect. Only fifty species are social parasites. Some, for instance the Yellow-billed Cuckoo *(Coccyzus americanus)*, generally rear their own broods and only now and then deposit an occasional egg in some other bird's nest. The remaining species, however, form pairs that breed normally, building nests and rearing their offspring jointly. Social parasitism is not limited to cuckoos. It is found among such birds as starlings, honeyguides and weavers and even some species of ducks.

Features common to the cuckoos and their immediate relatives are their grasping feet with two toes pointing forwards and two backwards, like the parrots, and the markedly long, graduated tail. Apart from one exception they are birds of solitary habits. Many species have a similar call. They have an unusual fondness for unpalatable foods other birds avoid (e.g. fuzzy, hairy caterpillars).

The Short-winged or **Indian Cuckoo** (273) is a species found in eastern Asia which, like the Common Cuckoo, deposits its eggs in the nests of small songsters.
Cuculus micropterus

272

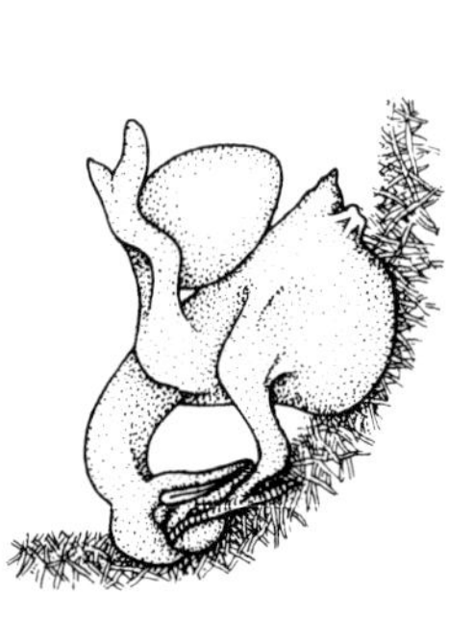

A fledgling cuckoo ejecting the eggs of its foster parents from the nest

273

274

The Common Cuckoo (272, 274), widespread throughout Europe, Asia and Africa, has many features that are adapted to its parasitic behaviour. The eggs are small and come in about twenty different colour variations that resemble the eggs of the host species; they range from uniform blue eggs such as the European Redstart's to greatly speckled eggs such as the Reed Warbler's. Because each cuckoo as a rule deposits its eggs in the nest of the species by which it was reared, it is possible to speak of 'biological' races, e.g. wagtail cuckoos, redstart cuckoos, shrike cuckoos, and the like. In some species of cuckoos (e.g. the Great Spotted Cuckoo – *Clamator glandarius*) that lay their eggs in the nests of larger birds and whose young are reared together with the hosts' offspring, adaptation is developed to such a degree that the fledgling cuckoos resemble the young of their foster parents for a certain period upon hatching. In spite of all these adaptations, however, many birds abandon their nest as soon as it is visited by a cuckoo or else stop feeding the fledgling. Since losses are by no means negligible the cuckoo lays as many as twenty-five eggs during the breeding season.

Cuculus canorus

275

The White-cheeked or **White-eared Turaco** (275) belongs to the family Musophagidae which numbers eighteen species of birds the size of a magpie to a raven and is found only on the African continent. Besides being related to the cuckoos it also has distant ties with certain gallinaceous birds, chiefly the hoatzin. The White-cheeked Turaco inhabits the forests of Ethiopia and Somaliland, keeping to the treetops where, like all members of this family, it feeds mostly on plant food – various fruits, buds and shoots. During the courtship display, performed to the sound of loud cries, the male spreads his wings wide to show off his beautiful red flight feathers. The red colour is produced by a pigment called turacin which contains free copper that dyes even the water in which the birds bathe. The nest is a simple, flimsy structure built in the branches. The clutch usually consists of two, dirty yellow, round eggs. The young, covered with a thick, dark coat of down on hatching, are fed by both parents. When they fledge the family remains together for a long time. The White-cheeked Turaco lives either solitary or in small groups. It has an unpleasant croaking voice.

Tauraco leucotis

Chapter 20 NOCTURNAL HUNTERS – OWLS

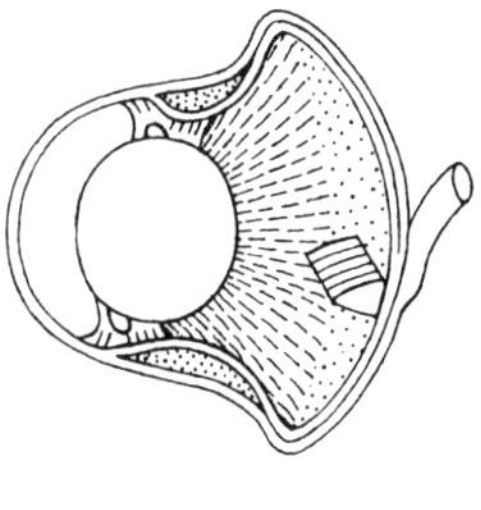

Cross-section of an owl's eye

The nocturnal habits of owls, their eerie and often frightening cries and the large, staring eyes have made these birds not only the object of great interest but also the subject of many superstitions and erroneous beliefs. These predators are able to hunt at night because of their keen sight and hearing. The large telescopic eyes with their excellent focusing ability, the spherical lens capable of catching and focusing even the slightest rays of light on the retina with its enormous number of light-sensitive cells, are the most highly developed in all the animal kingdom. They enable owls to see even in practically total darkness as well as in fog. Set facing forward they permit stereoscopic, three-dimensional vision as in man. The auditory organ is extremely sensitive and therefore protected against the pressure of strong sound waves by flaps of skin which partially cover the ear openings. The owl's hearing is further aided by the facial disc of fine feathers which functions as a sound reflector and enables the birds to pick up very slight sounds. The ear openings are located behind the eyes; the paired tufts of feathers on the head, which are often referred to as ears, are merely an ornament. The soft and fluffy plumage, which renders their flight so silent and enables them to capture their prey

276

277

unaware, is another feature of their superbly adapted equipment. Most owls feed primarily on small rodents, their numbers often greatly affecting the owls' nesting habits. Owls form a uniform group of about 140 species divided into two families.

The Barn Owl (276), typical inhabitant of church steeples and barn lofts, is widespread throughout Europe (except for Scandinavia), southern Asia, the whole of Africa, the warm parts of North America, South America and Australia. In years when the mice population is plentiful it sometimes rears more than ten youngsters, the second brood often fledging in late autumn. It feeds mostly on fieldmice. Like most owls it is resident; only young birds roam the countryside, as do older ones when food is scarce. In severe winters many owls die of hunger.
Tyto alba

The Tawny Owl (277) is a common species throughout Europe, except the northern part, in a broad belt across central Asia and North Africa in the Atlas Mountain countries. It is found both in lowland and mountain forests as well as city parks and gardens. It nests very early, and its hooting and excited cries can be heard the very first days that herald spring. The two to four, round, white eggs are usually laid in a tree hole. This owl feeds on mammals the size of a brown rat or smaller; some individuals hunt birds and even fish.
Strix aluco

The Great Grey Owl (278) is the Tawny Owl's counterpart in Scandinavia and northern Asia, and is also found in the colder parts of North America. It inhabits coniferous forests where it nests in abandoned raptors' nests.
Strix nebulosa

278

280

281

The Burrowing Owl *Speotyto cunicularia* (279), a comical, long-legged owl greatly resembling the Little Owl, inhabits the prairie lands of the southwestern United States and also South America. It nests in abandoned mammals' burrows. It can run very fast, which stands it in good stead when hunting prey on the ground – small mammals as well as lizards, snakes, frogs and insects; it also captures them in flight close above the ground.

The Tengmalm's Owl *Aegolius funereus* (280) is a totally nocturnal bird found in large coniferous and mixed forests in the boreal zone of Europe, Asia and North America; in central Europe it occurs only in the mountains. It is fond of nesting in abandoned Black Woodpeckers' holes. As with all owls, only the female incubates. After the young have hatched the male keeps the family supplied with food, delivering it to the female from outside the nest.

282

283

284

The Little Owl (281) is found near human habitations, in groves of old trees and in gardens.
Athene noctua
Besides small vertebrates it is also fond of insects and worms. It often starts hunting while it is still light. Its range includes temperate Eurasia as far as the Amur River and Korea, also North Africa.

The Cuban Pygmy Owl (282) is one of eleven species of very small owls that are found in forests
Glaucidium siju
throughout the world except Australia. The Cuban Pygmy Owl, scarcely as large as a starling, occurs only in Cuba. Pygmy owls have many features in common. They are all active predators that often hunt even during the day, their chief victims being small mammals and birds. They lay by stores of food during the nesting season and in periods of inclement weather. They are active birds which have the habit of jerking their tails from side to side when upset. The calls range from a monotonously repeated whistle that sounds like 'diub' to loud barking cries when the birds are upset. It is fond of nesting in abandoned woodpeckers' holes. While rearing the young the female keeps the hole spotlessly clean, carefully removing all remnants of food, which is not a common habit amongst owls. The young are fed from the store of food every few hours, even during the day. They leave the nest while still unable to fly, remaining in its vicinity for a number of days.

The African Wood Owl (283) is the most common forest owl in the whole of Africa south of the
Ciccaba woodfordii
Sahara. It inhabits the virgin forests alongside rivers as well as dense, evergreen groves if there is water nearby. In the mountains it occurs at elevations up to and even above two thousand metres. Besides small mammals it is fond of catching large beetles on the wing.

The Ural Owl (284) ranges from eastern Europe across the whole wooded part of Asia all the
Strix uralensis
way to Japan. In central Europe it has a more continuous distribution

285

286

in the Carpathians and nests here and there in the eastern Alps. It looks like a bigger version of the Tawny Owl but has a longer, rounded tail. Like the Tawny Owl it occurs in two colour phases – grey and dark brown. It inhabits silent, broadleaved and mixed woods, where it nests in tree holes or in abandoned raptors' nests. The Ural Owl feeds mainly on small rodents and the number of nesting birds varies in accordance with the size of the rodent population in the given year.

The Brown Fish Owl (285) is one of four species of owls, closely related to the eagle owls, which feed mainly on fish. For that reason they have only slightly feathered legs and naked toes, which have rough scales on the underside that help the bird hold the slippery prey. The Brown Fish Owl is almost as big as the Eurasian Eagle Owl, is a uniform brown with fine streaks, has an imperfect facial disc and long ear-tufts jutting out at the sides. It is at home in southern and eastern Asia from Syria and Iraq across India and southern China north to the Sea of Okhotsk. It hunts its prey, which besides fish includes also crayfish, crabs and frogs, in streams and river shallows either by wading in the water or lying in wait. It hunts in the evening and at night. It nests in various tree holes, rock cavities as well as in treetops in abandoned open raptors' nests. The clutch consists of one or two white eggs.

Ketupa zeylonensis

The Great Horned Owl (286) is an American species ranging from the forest belt of northern Canada to the Strait of Magellan. There are several regional variations differing primarily in colouring. It generally rears its two to three young in the abandoned nests of raptors. The diet consists of all sorts of small vertebrates, but it has been known to attack even skunks and porcupines.

Bubo virginianus

The Snowy Owl (287) has a circumpolar arctic distribution, central Scandinavia marking the extreme southern boundary of its range in Europe. It breeds in the treeless tundras and like other northern owls is a bird of diurnal habit. It nests on the bare ground. Lemmings are the mainstay of the diet and in years when these are scarce the Snowy Owl either does not nest at all or lays only three or four eggs, whereas in years when they are plentiful it lays twice that number — six to nine. Some birds fly farther south for the winter; sometimes these flights are in the form of invasions reaching as far south as central and western Europe.

Nyctea scandiaca

The Eurasian Eagle Owl (288, 289), largest of all living owls, is found throughout most of Asia, Europe and North Africa. Persecuted by man, it has become extinct in many central European countries. It is found in wooded country in both mountains and level land, its favourite nesting site being rock ledges though it also nests in abandoned raptors' nests. The birds remain paired for life and nest in the same spot for many years. In mild winters the Eagle Owl begins nesting very early, sometimes even in mid-February. The clutch consists of two to four round white eggs with a grainy surface. Only the female incubates while the food is brought by the male. His arrival is announced by the characteristic call 'boo-hoo, boo-hoo', to which the female responds with a piercing shriek that sounds like 'khrank'. The Eagle Owl's diet consists mainly of fieldmice, brown rats, squirrels, hedgehogs and wild rabbits. In lake and pond country the Eagle Owl hunts waterfowl, grebes, coots and ducks. It hunts only at night. As it is becoming an uncommon species it deserves full protection by law.

Bubo bubo

287

288

289

290

291

The Screech Owl (290) is one of the 22 species and many races of small, eared owls with yellow eyes found throughout the world excepting Australia. It has two colour phases (or varieties) – grey and reddish-brown – and inhabits North America from southern Alaska and Canada to Mexico and Florida. It is fond of nesting in woodpecker holes as well as man-made nestboxes, even near human habitations. The diet consists almost entirely of insects.
Otus asio

The Long-eared Owl (291), one of the commonest owls throughout the temperate zone of Europe, Asia and North America, is generally found in areas with scattered woods set amidst spreading fields and meadows. It does not build a nest of its own, preferring to avail itself of an abandoned crow's, magpie's or squirrel's nest. The diet consists almost solely of fieldmice. In years when these are plentiful the Long-eared Owl often has a second brood in late summer. The clutch consists of four to seven white eggs incubated by the female while the male keeps guard close by. The owls hunt at dusk and during the hours of darkness, in winter even in the daytime. Birds that breed in central Europe fly to western Europe in the autumn; the northeastern populations migrate to central Europe.
Asio otus

The Short-eared Owl (292, 293) is widely distributed in Europe, Asia, North and South America. In the north it inhabits the tundra, in more southerly latitudes it is found on wet meadows, fallow land and semi-steppes. The Short-eared Owl always nests on the ground and is the only owl to gather material (various plant parts) for the nest. It changes its nesting site according to the abundance of fieldmice. The clutch comprises eight to ten eggs, which the female begins incubating as soon as the first is laid. The nest is surrounded by a large store of fieldmice. The Short-eared Owl is commonly active even in the daytime. Northern populations migrate, but not regularly, as far south as the Mediterranean region and southern Asia.
Asio flammeus

292

293

Chapter 21 MORE NOCTURNAL BIRDS

Sensing danger the European Nightjar hovers in one spot in the air and looks around

In speaking of nocturnal birds one always thinks of owls, forgetting that there are many more that take to the wing at dusk or are active at night, chief among them being the goatsuckers. The members of this large order, comprising five families and eighty-six species, are to be found in all parts of the world, primarily in the tropical zone. Some species, however, occur also in the temperate zone but there are no goatsuckers in the far north. Goatsuckers resemble owls in a number of respects. They are nocturnal birds, have large eyes and soft, fluffy plumage, coloured a sobre grey or brownish-grey with fine darker markings, which blends well with the bark of the trees where the birds rest during the day, thus making them practically invisible. Other characteristics are reminiscent of another group of birds which, like the goatsuckers, catch insects on the wing, namely swifts. Goatsuckers have a flat head, long, narrow wings, small bill and tremendously wide gape. Growing at the base of the bill are long bristles that serve as a trap for catching insects in the dark, thus increasing the effectiveness of their method of securing food. Like the swifts, goatsuckers also have short legs and small toes. Many species sit lengthwise on branches, not crosswise as most birds, for the small toes are incapable of grasping thicker branches. Both parents take turns incubating the eggs for a comparatively short time — sixteen to twenty-one days. The young hatch seeing and covered with a coat of mottled down, and within a few days run around the nest on their short legs.

The largest family of the order of goatsuckers (Caprimulgiformes) is the nominate group known as nightjars (Caprimulgidae), numbering sixty-eight species, most of which live in the tropics.

The Whip-poor-will *Caprimulgus vociferus* (294), whose name accords with its call, is at home in the forests of eastern North America. On a quiet summer evening its note can be heard repeated at one-second intervals sometimes as much as several hundred times in succession. It is the male establishing his nesting territory.

294

295

The European Nightjar (295) is a woodland bird of the temperate zone of Asia and Europe, its range extending north as far as Norway and southern Sweden. It is partial to warm, dry forests in the lowlands. It is a migratory bird which returns to its breeding grounds from its winter quarters in Africa fairly late in the spring. Immediately after its arrival the weird call of the males pierces the hush of the twilight hours — a persistent, monotonous note alternately pitched high and then low. While singing his courting song the male sits lengthwise on a thick branch. Then flying from one branch to another he performs the next stage of his courtship in the air — clapping his narrow wings together above his back and uttering a churring note that is difficult to describe. The two, grey mottled eggs are laid in a shallow, unlined scrape in the ground, their coloration making them inconspicuous among the litter. The male takes little part in incubating. His worries begin after about eighteen days when the young hatch is covered with dark powder down. When feeding the young flying insects the beaks of the parents are engulfed by those of their offspring. Nightjars sometimes have two broods a year. The adults ward off intruders from the nest by ruffling their feathers and making a hissing sound or else try to draw attention away from the young by pretending to be wounded. During long spells of harsh weather nightjars fall into a torpid state, what might be termed hibernation, during which the metabolism is lowered and the body temperature slowly drops almost to the level of the ambient temperature.

Caprimulgus europaeus

The Jungle Nightjar (296) differs from the preceding species by being somewhat larger and more darkly coloured. It occurs in southeastern and southern Asia, being quite plentiful in places at the edges of woods. It often flies into villages in search of food.

Caprimulgus indicus

296

The Oilbird or **Guacharo** (297) does not have as fine feathers as the other goatsuckers and differs in biological respects as well, this being the reason why it has been placed in a separate family (Steatornithidae). The Oilbird does not feed on animal food but on fruits. It occurs in northern South America in deep caves, building its large nest of droppings mixed with palm seeds on rocky ledges inside the caves. The young, two to four, remain in the nest in the dark cave for three to four months. On a diet of oily fruits they are so fat by the time they are seventy days old that they weigh

Steatornis caripensis

297

one-and-a-half times as much as their parents. It is then that the Indians take them out of the nest and extract cooking oil from them, the layer of fat on their bellies being exceptionally thick. Nesting colonies are found in the darkness of long underground passages sometimes measuring as much as several hundred metres. Oilbirds, like bats, emit a high, short sound, though it is not in the ultrasonic range, which serves as an echo-ranger aiding the bird in its flight. If disturbed by man, their nesting colonies sound a deafening cry — hence their original Indian name 'Guacharo' (= bawler).

The Tawny Frogmouth (298), the best known of the frogmouths (family Podargidae), is a native of Australia and Tasmania. It catches food on the wing only occasionally, preying mostly on insects in the treetops. Sometimes it even catches large insects and small vertebrates on the ground. During the day it rests motionless in the treetops, coming to life only at dusk. At this time, occasionally also throughout the night, it repeatedly utters its grunting two-syllable note. The Tawny Frogmouth has short, blunt wings and is not a particularly agile flier.

Podargus strigoides

298

Chapter 22 EXCEPTIONAL FLIERS

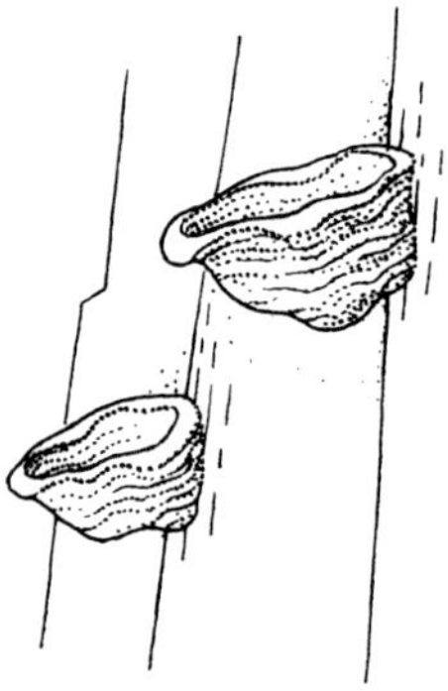

Swiftlet's nest of hardened saliva

The flight performances of birds may be judged by a number of different criteria. Except for 'motorless' gliding flight and long-distance flights the record for speed, agility and endurance is held by the swifts of the order Apodiformes, which also includes the very different family of hummingbirds. Though they differ in the way they feed and in their nesting habits, the two families have many anatomical features in common. The most conspicuous is the wing structure with its short scapula and extremely long primaries. However, the method of flight is entirely different. The legs are short and the leg muscles weak, and neither the hummingbirds nor the swifts use them for running.

The swifts and crested swifts (family Apodidae and Hemiprocnidae) are exceptionally fast fliers. The wings are long and narrow and curve back in a crescent. The wing quills are the longest of all birds in proportion to the body size. A striking feature is the tiny beak with wide gape, reminiscent of the nightjar's, with which the birds catch insects on the wing. Swifts have large salivary glands which become even bigger during the nesting season and produce a sticky saliva that hardens in the air and is used by the birds to glue their nests. The one to six eggs are incubated by both parents and both share the duties of feeding. Swifts are soberly coloured in shades of grey, black and brown. The seventy-seven species are practically worldwide in distribution; only one occurs within the Arctic Circle.

The Common Swift (299) is so superbly adapted to life in the air that it not only finds food for itself and its offspring in the air but also catches nesting material and even mates on the wing. The short leg with four toes pointing forward serves only for clinging to cliffs, cornices and tree trunks. Originally the Swift nested mostly on cliffs and in trees but today most of these birds prefer to nest in human dwellings. The nest of feathers and plant down is glued together with saliva. The two to three eggs are pure white and the young hatch after eighteen days. During long spells of unfavourable weather, when the food supply is inadequate, swifts fall into a torpid state, during which their body temperature drops and body metabolism slows. This occurs more often in the case of young birds, for adults, with their fast flight (up to 160 kilometres per hour) and endurance, are able to seek food in distant places. The Common Swift is widely distributed throughout Europe, northwest Africa and Asia as far as Manchuria. European populations winter in Africa south of the Sahara, remaining in their nesting territories only three months or so.

Apus apus

The speediest fliers are to be found amongst the so-called spinetailed swifts which take their name from the long tail-feather shafts projecting beyond the vanes as needles and serving as a prop on cliffs and tree trunks. This group includes fifty species found in the tropics.

The White-throated Spinetailed Swift (300) breeds in eastern and southern Asia. The nest, a mixture of moss and hairs glued together with saliva, is placed in tree holes; the three to seven eggs are sometimes laid directly on the floor of the cavity.

Chaetura caudacuta

The swift family also includes the swiftlets of southern Asia and Indonesia which glue their nests of hardened saliva to the walls of seacoast cliffs. In some countries the saliva nests are used for soup, and are said to have an invigorating effect.

300

301

302

Wealth of hummingbird species in the New World

Hummingbirds (Trochilidae) are an endemic group found in the New World. More than three hundred species have been described, most of them inhabitants of the tropical zone of South America. The hummingbird family includes both large and tiny species — the largest measures twenty-one centimetres (the Giant Hummingbird); the smallest — the Bee Hummingbirds (genus *Calypte*), measuring 5.5 centimetres and weighing less than two grams, are the smallest birds known. They can flap their wings so rapidly (50—70 wing-beats per second) that the human eye sees only an indistinct blur. The action of their wings permits

303

304

hummingbirds to change the direction of their flight with amazing speed and to fly in any direction, including backwards (they are the only birds able to do so). Hovering in one spot they suck nectar from flowers with their specially adapted tubular tongues, at the same time pollinating the tropical plants. Many species have vivid iridescent colours and various feather ornaments.

The Ruby-throated Hummingbird (301) ranges from Florida and Texas all the way to southern Labrador and Lake Athabasca. It winters south of Texas and on the islands of the Caribbean. Remarkable is the performance of this eight-centimetre bird which when migrating flies as much as 800 kilometres over the sea.
Archilochus colubris

The Sparkling Violet-ear (302) inhabits the virgin forests and shrubby plains of northwestern South America.
Colibri coruscans

The Broad-tailed Hummingbird (303) breeds in the mountain regions of the southern United States and in Guatemala. The female, as in all other species, lays two white eggs in the neat nest of moss, plant down and spiders' webs. The males are poor husbands in that they generally take no part in building the nest or caring for the young.
Selasphorus platycercus

The Golden-bellied Hummingbird (304) is one of eleven small species of the genus *Chlorostilbon*. It has beautiful, shining green plumage with puffy feather ornaments on the flanks. Its bill is slightly widened at the base. It inhabits northern Argentina, Bolivia and Paraguay.
Chlorostilbon aureoventris

Chapter 23 COLOURFUL BEAUTIES

Combinations of all kinds of bright and sober colours in a wide range of hues are to be found also amongst the birds of the large order Coraciiformes. Though otherwise extremely diversified they have several anatomical features in common, most striking being the fusing of the front toes for part of their length. All are furthermore fond of nesting in cavities, either in trees or in burrows in the ground which they dig for themselves. The young, excepting those of the Hoopoe, are born naked. The seven families, embracing some 190 species, are mostly tropical and subtropical birds with a cosmopolitan distribution.

The stout and strong bill of the rollers (family Coraciidae) resembles that of the crows. The prevailing colours of the seventeen Old World species are blue, green and brown.

The European Roller *Coracias garrulus*

The European Roller (305), coloured black, chestnut and blue, inhabits the temperate lands of Eurasia and Morocco. It nests either in abandoned woodpeckers' holes in trees or in other cavities in sand and mud banks. The courtship display of the male is an amazing sight, he rolls and wheels through the air with an occasional somersault accompanied by loud, harsh cries. The female lays four or five white eggs. The birds feed on insects and small vertebrates. They winter in East and South Africa.

The elegant, tastefully coloured and slender bee-eaters (Meropidae) are among the most attractive of birds. They are at home in the warm forest-steppes and steppes of the Old World. They resemble large swallows in shape and flight. The tail, however, is not forked. On the

305

Courtship display of the European Bee-eater

306

contrary, the two central tail feathers are considerably elongated. The family numbers twenty-five species.

The European Bee-eater (306), a magnificently coloured bird with green, blue, yellow, black and white plumage, brightens the warm regions of southern Europe, the Middle East, southwestern Asia and North Africa. Its existence depends on a sufficient supply of hymenopterous insects and also mud or sand

Merops apiaster

307

banks in which to dig its nest – a burrow up to two metres long, excavated by both partners. Unlike the rollers, bee-eaters are social birds that often form small colonies.

The Madagascar or **Blue-cheeked Bee-eater** (307) is a glowing green colour with black eye stripe edged with blue. The throat and underside of the wings are rufous red. It is widely distributed in the steppes of southern Asia and northern Africa. The nesting site is a burrow in the ground. Its habits are similar to the European Bee-eater's.

Merops superciliosus

The kingfishers (Alcedinidae) are distinguished by a thick-set body, large head, strong bill and short legs. The large bill is a good instrument for catching fish, which along with insects form the bulk of the kingfishers' diet. The nesting burrows, which the birds usually dig themselves, end in a round chamber where the pure-white eggs are laid. The male and female, both coloured alike, share the duties of incubating and rearing the young. The family of kingfishers numbers eighty-seven species found in all parts of the world.

The Common Kingfisher (308, 309), a bird of clear, running watercourses, is an expert fisher. It perches motionless for long periods of time on a branch overhanging the water, plunging headlong after a fish as soon as it spots one. It kills the fish by pounding it against a branch and then swallows it head first. When offering a fish to the female, which is part of the courtship performance, or when feeding it to the young the fish is turned the other way round as shown in the picture. The nesting burrow is usually dug in banks along watercourses. The burrow, a 60 to 100-centimetre-long tunnel ending in a roomy nesting chamber, is dug by both partners. It is not lined but in time the undigested fish bones and scales with which it becomes littered form a soft pad. The Kingfisher has one to three broods a year; the clutch consists of six or seven eggs. At first the young are fed small fish and insects, later larger ones. They leave the nest after twenty-two to twenty-seven days to fend for themselves, sometimes roaming far from the

Alcedo atthis

308

30

310

311

natal site. The Common Kingfisher is at home throughout the temperate zone of Europe and Asia, its range extending south as far as Sri Lanka, Sumatra and the Solomons.

The largest group of tropical kingfishers are those of the genus *Halcyon*, which includes some 150 forms. In the main they are medium-size birds with black, red or, in some instances, bi-coloured beaks.

The Blue-breasted Kingfisher (310) is one of the common species found in Africa south of the Sahara.
Halcyon malimbicus

The White-breasted Kingfisher (311) is the best known of the group. The plumage is predominantly brown, the back, tail and wings are coloured blue and the bill and legs a bright red. It ranges from Asia Minor to Taiwan, Vietnam and the Philippines. It is not tied to water and is often found even in contiguous virgin forests, for besides fish it also feeds on land vertebrates, larger insects and crustaceans.
Halcyon smyrnensis

The Kookaburra or **Laughing Jackass** (312), of southeastern Australia, is the biggest of the kingfishers. It is coloured a sober brown, black and white, has a short crest on the nape, a large beak, and a tail that is comparatively long for a kingfisher. It is a solitary bird and is not confined to water, occurring in any type of country with trees and shrubs where it can hunt its food. It is also found in city parks and gardens. The diet, from which fish are absent, consists of vertebrates which the bird kills adroitly with its strong beak. It is fond of snakes, particularly poisonous snakes, which is the reason for its popularity. It nests in tree cavities or termite nests. The two to four young hatch from the pure white eggs in September to December after an incubation period of twenty-five days. The Kookaburra is very aggressive in its defence of the young during the rearing period and is capable of causing painful wounds even to a human if he approaches the nest. Otherwise it is an intriguing and inquisitive bird, which is very popular with Australians and is even regarded as that country's national bird. Its only unpleasant characteristic is its piercing, weird, laughing cry.
Dacelo gigas

Range of motmots

A distinct New World family are the motmots (Momotidae). The eight species, coloured mostly blue, green and brown, have much in common with the rollers. They capture their food both in the air and on the ground, feeding mainly on large insects, but eating also smaller vertebrates and sometimes even fruit. They nest in burrows dug in vertical banks as well as in flat ground and sometimes in rock crevices. The three to four white eggs are incubated by both partners for about three weeks and both feed the young a further four to five weeks. Motmots have a long tail which they twitch from side to side when excited, especially when courting. The two central tail feathers are elongated and the vanes at the end break or wear away in time leaving the shafts bare and giving the tail its characteristic racquet-tipped look. However, when the bird moults the new feathers grow in fully webbed, the vanes at the end again breaking off as the bird preens them. Motmots inhabit the forests of Central and South America.

The Blue-crowned Motmot or **King-of-the-woods** (313) inhabits Central America and the northern parts of South America. It nests either in a burrow it digs itself or in the abandoned burrow of mammals where it digs side tunnels. The motmots' nesting tunnels are sometimes as much as three metres long, often twisting and winding, and they terminate in a large nesting chamber where the eggs are laid on the bare earth. Usually the tunnels are excavated in vertical banks alongside rivers or pathways. It is interesting to note that motmots begin digging a new tunnel shortly after the young have fledged, several months before they start laying eggs, but do not use it for sleeping quarters. The adult birds take turns incubating for twenty-two days. The young fledglings are strikingly like their parents except that they do not have the small fans at the tip of the tail.

Momotus momota

The Hoopoe (314), the sole member of the family Upupidae, differs from its rainbow-coloured relatives in that its plumage is a subdued orange-brown with black and white pattern. On the other hand it sports a flat ornamental crest of long feathers resting folded on the nape except when the bird is excited, when they are raised and expanded to form a magnificent fan. The Hoopoe

Upupa epops

312

313

314

315

inhabits open country and is found in a large part of Eurasia, Africa as well as in Madagascar. It nests in holes and semi-cavities. The young birds provoke the parents into feeding them by opening their beaks wide to expose the bright red throats, as do songbirds. During the nesting period the young as well as the female exude a foul-smelling liquid from the preen gland which they are able to eject at intruders. The Hoopoe has a very short tongue in its long bill; that is the reason for its swallowing food – mostly grasshoppers, locusts, crickets and various larvae – by throwing it up to the air and catching it back right into the gullet.

The Quetzal (315), another magnificent gem which we have included in this list of colourful beauties, belongs to the order of trogons (Trogoniformes), the only birds with the first and second, not the first and fourth, toes reversed. The thirty-four species of trogons are found in the tropics of the Old and New World. The Quetzal is the best known of all mainly because of its beautiful and magnificently coloured plumage. On the head it sports a short crest, the long wing coverts are like the fronds of a palm leaf and the two upper tail feathers grow into gorgeous plumes up to one metre long. The Quetzal nests in tree cavities in forests from southern Mexico to Costa Rica. It appears on the Guatemalan state seal.

Pharomachrus mocino

Chapter 24 BIRDS THAT SEAL THEIR MATES IN THE NEST

A male hornbill feeds his sealed-in mate

The large order Coraciiformes, birds characterized by having their front toes joined for part of their length, includes the well-known family of hornbills (Bucerotidae) which have enormous, peculiarly shaped bills and sober plumage. Besides this they have interesting nesting habits such as sealing the entrance to the cavity in which the female incubates, a characteristic typical of the forty-five species that make up this family. Apart from two exceptions they are all arboreal birds inhabiting the forested areas of Africa, southern Arabia, India and southeast Asia. For rearing their young, hornbills seek out old trees with suitable cavities which the partners seal with a combination of mud, dirt, their own excrement and bits of vegetation, cemented together either with their droppings or matter regurgitated from the crop. When the opening is so narrow that the female can just squeeze through, the male continues to plaster it up from the outside and the female from within, leaving only a narrow slit for the bill. The attentive male feeds his mate throughout the entire period of incubation and for a long time after the young have hatched. The length of time the female remains walled in varies according to the different species. In some the female pecks her way out after a few weeks to help her mate feed the young, which seal themselves in again; in other species she leaves the cavity together with her offspring. It is interesting to note that as soon as she is sealed in the female begins to shed her tail and wing quills as well as contour feathers which renders her incapable of flight. The moult is apparently triggered by the stay in darkness for females that do not nest moult successively and do not lose their power of flight. The male's performance in feeding his mate and offspring is truly remarkable. A male of the species *Bycanistes brevis* is recorded as having made 1,660 trips to the nest over a period of seventeen days, bringing 24,000 fruits.

The Abyssinian Ground Hornbill (316) nests in tree holes but spends most of its time on the ground hunting vertebrates and gathering fruits. Whereas other hornbills hop about in an ungainly manner the Abyssinian Ground Hornbill moves with a deliberate gait. The toes are only slightly joined. It does not seal the entrance to the cavity. The female sheds her feathers successively and thus does not lose her power of flight. This species inhabits the African savannas.
Bucorvus abyssinicus

The Red-billed Hornbill (317) does not have the casque on the bill so that it slightly resembles a toucan. It nests in Africa south of Somaliland and Senegal. The female lays three to five eggs which she incubates for three weeks, remaining sealed in with her offspring a further three weeks. It takes about five hours for her to chip away the hard, brick-like wall so that she can leave the nest. After this the young seal themselves in again with droppings mixed with food remnants and the parents continue to feed them. They chip their way out after about six weeks.
Tockus erythrorhynchus

The Indian Pied Hornbill (318) has a tube-shaped casque on top of the bill, which in the male extends into a point in front. In small groups it seeks various fruits in the treetops and enlivens the forests of India and Malaysia with its presence.
Anthracoceros malabaricus

The Blyth's Hornbill (319), a large hornbill about one metre long, inhabits the islands of the Pacific from the Moluccas to the Solomon Islands. Besides fruit it feeds also on various invertebrates and vertebrates, such as lizards, snakes as well as small mammals and birds.
Rhyticeros plicatus

The Great Hornbill (320, 321) is found in the forests of southeast Asia ranging from India to the Himalayas, the Malay Peninsula and Sumatra. Its bill, almost twenty-
Buceros bicornis

317

318

319

320

321

five centimetres long, is truly remarkable. In males the casque is deeply grooved above and two parallel projections extend far onto the head. In the female the casque is flat. The Great Hornbill keeps mainly to the treetops. It feeds on various fruits, being partial to figs, but will swallow any animal it can handle. It nests in tree cavities, which in view of its size (1.2 metres in length) must be quite large. The same cavity is often used for a number of years. It is believed to rear only one offspring. The female leaves the nesting cavity a month before the young bird fledges and the latter seals itself in again, continuing to be fed by both parents.

Chapter 25

MASTER CARPENTERS AND THEIR KIN

Not all members of the order Piciformes climb trees, drill holes in wood to excavate insect larvae and chisel cavities in tree trunks. Such uniformity cannot be expected among the 383 species that make up the six families of this order. Despite this, however, the various families have many features in common, both in body structure and in the arrangement of the feathers. A conspicuous feature is the similarity in the structure of the foot with the first and fourth toes pointing backwards as in the parrots, which likewise show a great aptitude for climbing. In general it may be said of the birds of this order that they are all arboreal and nest in cavities, their eggs are white and the young hatch naked and blind. With the exception of Australia, Madagascar, New Zealand and the islands of the south seas they have a worldwide distribution.

The barbets (family Capitonidae) are all gaudily coloured birds which have conspicuous bristles at the base of the bill that give them a bearded look – hence their name. They inhabit the tropical forests of America, Africa and Asia. Barbets are rarely seen on the ground. They gather their food, both animal and vegetable, in branches of trees, sometimes climbing trunks like woodpeckers. They lay their eggs in nesting holes which they excavate themselves in decaying wood.

The Black-backed or **Levaillant's Barbet** (322) and its closest relatives are not confined to trees like other barbets. It is often seen on the ground. It inhabits the bushlands of eastern and southern Africa.

Trachyphonus vaillantii

322

323

The Wryneck (323) *Jynx torquilla* The above general description does not apply in all details to Wrynecks. They have not developed the structural features of the truly perfect type of climbing birds. The tail feathers are not stiff and the bill is small and incapable of chiselling. A favourite food of the Wryneck are ants which it gathers on the ground. When danger threatens it turns its head at queer angles and erects its head crest. The Wryneck breeds in Europe and Asia as far as Japan.

The Great Hill Barbet (324) *Megalaima virens* is one of the larger species – about the size of a Jackdaw. It feeds on fruits and flowers. It is at home in southern Asia, its range extending from Kashmir to southern China and Vietnam.

The true master carpenters are what one may call the birds of the woodpecker family (Picidae). Most are structurally adapted for climbing round tree trunks, particularly their foot which has two toes pointing forward and two backward and is equipped with strong claws. An excellent carpenter's tool is the strong, laterally compressed bill, the straight tip either blunt or pointed, which is used like a chisel or an axe.

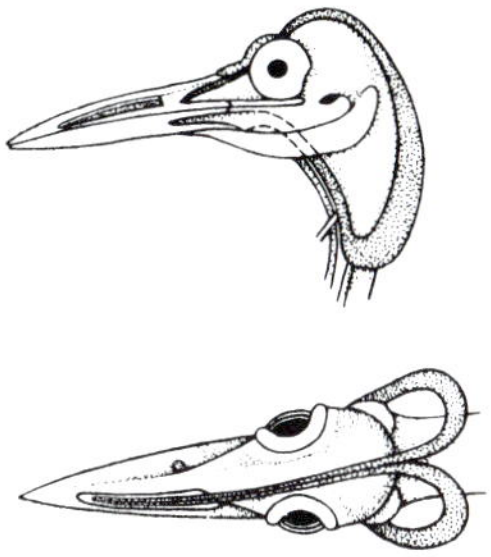

The long extendible tongue of the woodpeckers is manipulated by the rear horns of the hyoid, coiled in a loop around the skull

324 325

With their bills woodpeckers probe under bark or drill in wood to get at the insects concealed there and also excavate large nesting holes. Their feet alone would not provide sufficient support for such work, requiring fairly large swings of the head as the bird perches vertically on the trunk, and that is where the strong, wedge-shaped tail comes in with its long, pointed feathers with broad, stiff rachis which function like a flexible spring and serve as a prop. The woodpecker's tongue is likewise specially adapted for extracting insects from narrow cracks. It is in fact a long, extendible harpoon whose hard, pointed tip, equipped with barbs and coated with a gluey substance, hooks the prey. The prolonged horns of the hyoid, powered by special muscles, enable the woodpecker to stick out the tongue to astonishing lengths beyond the bill. Not all woodpeckers, however, feed on insects. Some drink the sap of trees, others eat a combined diet of animal and vegetable food. The 209 species of woodpeckers are to be found throughout the world excepting the polar regions, Madagascar, Australia, New Guinea and neighbouring islands.

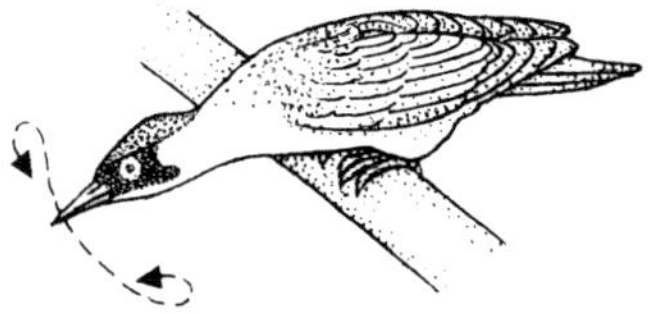

Courtship display of the Green Woodpecker *(Picus viridis)*

326

327

The Great Spotted Woodpecker (325, 326) reveals its presence in spring by its rapid drumming on resonant tree trunks or stumps of branches, the same as many of its relatives. This drumming is a substitute for song and serves to establish the bird's nesting territory. It is at home in Europe and Asia, except for southern Asia.
Dendrocopos major

The Lesser Spotted Woodpecker (327) has almost the same range of distribution as the Great Spotted Woodpecker, both species often occurring in the same habitat. The Lesser Spotted Woodpecker, however, shows a preference for deciduous and mixed woods and is not found at high elevations. It presents no competition to larger woodpeckers as far as food is concerned for unlike the latter it stays well up among the branches of trees rather than on the trunks.
Dendrocopos minor

The Pileated Woodpecker (328) is coloured black and white. The male has a large red crest and red whisker mark; the female a small red crest and black whisker mark. This woodpecker is widespread in the forests of North America; it is most plentiful in swampy woodlands.
Dryocopus pileatus

The Common Flicker (329) is the most brightly-coloured of the woodpeckers. It excavates nesting cavities in tree trunks, telegraph poles and construction beams. Like the Green Woodpecker it also often moves about on the ground looking for ants which it picks up with its sticky tongue.
Colaptes auratus

328

329

Chapter 26 BIRDS WITH LARGE BEAKS

Range of toucans

Though toucans (family Ramphastidae) belong to the same order as woodpeckers (Piciformes) they are dealt with separately in this book. They are a remarkably interesting family comprising thirty-seven vividly coloured species of birds distinguished mainly by their huge bills. Few birds have such marked body asymmetry. The tremendous laterally compressed bill appears to be only a cumbersome burden to its owner. In reality, however, it is very light, being hollow and reinforced with thin, horny lamellae. The surface is covered with a thin, horny layer that is generally brightly coloured in hues characteristic for the given species. One wonders, however, about the use of such a bill; for a smaller one would surely suffice for gathering and eating the large fruits the birds feed on. Perhaps the bright coloration plays a symbolic role in the birds' social life. Toucans are birds of the forest that nest in cavities. Both mates share the duties of nesting.

The Swainson's Toucan (330) is at home in the tropical and subtropical forests of Central *Ramphastos swainsonii* America from Honduras south to Ecuador and east to Venezuela. The slanting markings on the bill are not common in toucans.

The Toco Toucan (331), largest of all toucans, inhabits the central and northern regions of South *Ramphastos toco* America. It feeds on various fruits which it adroitly plucks with its bill and masticates before swallowing if they are too large. Since toucans in

330

332

333

captivity eat animal food it is believed they do so as well in the wild. They are even suspected of robbing the nests of other birds.

The Keel-billed Toucan (332) has a very brightly coloured green bill with red tip, pale orange patch on the upper bill and black band at the base. It inhabits the rainforests from Mexico to Venezuela.
Ramphastos sulfuratus

334

The Channel-billed Toucan (333), which has a black bill with grey band at the base, is more commonly known as the Ariel Toucan, though this name, as has been proved, applies only to the east Brazilian race that breeds in Venezuela, Guyana and Brazil north of the Amazon River.
Ramphastos vitellinus

The black toucans of the genus *Ramphastos* are all large birds with large bills. Toucans of the genus *Pteroglossus* are smaller and have smaller bills with a flat-edged upper bill. The beaks are not as brilliantly coloured as those of the black-bodied toucans, being whitish or black and white. The name *Pteroglossus* means 'feathered tongue'. However, this characteristic is not unique to this genus. All toucans have a long, narrow tongue fringed on both sides with bristles reminiscent of a feather.

The Collared or **Banded Aracari** (334) has a white upper bill with red patch at the base and black lower bill with white basal stripe. It occurs in several forms in Central America, Colombia and Venezuela.
Pteroglossus torquatus

Seven other species of small toucans – the green toucanets of the genus *Aulacorhynchus* – are also inhabitants of Central America and the northern parts of South America. However, they are mostly found at higher elevations, some species never descending to lower altitudes. They are very social birds and disperse to form pairs only prior to nesting, reforming into large or smaller groups as soon as the young can fend for themselves. They sometimes even fly in groups from place to place in search of food. Toucanets nest in natural holes as well as in abandoned woodpeckers' nests, which they adapt with their bills.

The Emerald Toucanet (335), a small toucan coloured olive-green with pale upper and black lower bill, inhabits the territory from Mexico to Peru. It may be called
Aulacorhynchus prasinus

335

336

a highland species for it is found at altitudes above 2,500 metres. As a cage bird it requires no special care for it is quite hardy.

The Crimson-rumped Toucanet (336) is mostly green, with red-and-black bill and crimson rump. As in the preceding species the incubating period is sixteen days and the young do not leave the nesting cavity until they are forty-three days old. It is not recommended to keep this toucanet in a cage together with smaller species for it sometimes kills and even eats them. The sexes are both coloured alike, the same as in other species. This toucanet is found in the Andes in Colombia and Ecuador.

Aulacorhynchus haematopygus

Nestling toucans do not rest on their short, weak legs but on prominent pads on their heels. These heel pads remain the whole time the young are in the nest; later they disappear. Similar pads are to be found in the young of birds of the woodpecker family. Another distinctive feature is the position adopted by toucans when they sleep. Perching on a branch they lay the bill down the centre of the back and cover it with the tail feathers, the tips of which reach as far as the shoulder so that the bird looks like a ball of feathers.

Chapter 27 SONGBIRDS THAT DO NOT SING

In most books about the birds of the world relatively little space is devoted to the largest order of birds — the songbirds or passerines (order Passeriformes). For example, in Brehm's classic animal encyclopedia *Brehms Tierleben* only one of the four volumes on birds deals with this order. If one considers that it contains about 5,100 species, which is more than all other bird orders combined, then the ratio should be reversed. There is no denying that the small passerines are less attractive than, for instance, birds of prey, owls or waterfowl, and that they include many species of like size and appearance. In the wild practically every birdwatcher will focus his fieldglasses on the large, even though uncommon, species, rather than on the small songsters which are to be found everywhere in great numbers. This chapter deals with several species of songbirds that are placed lower on the evolutionary scale whereas the last chapter is devoted solely to the suborder of true songbirds (Oscines). The birds in this chapter differ from the true songsters in that they have a simpler arrangement of the voice box, which has three or fewer pairs of syrinx muscles whereas the true Oscines have seven to nine pairs of syrinx muscles. The title of this chapter is not wholly accurate for some of the birds listed here include species that are fairly talented vocalists. In general, however, they are not able to match the performance of the true songbirds.

Passerines all have the same foot structure with the four toes joined at the same level and only the first pointing backwards. They brighten all types of environments and form large communities. All have nidicolous young.

The Lesser Green Broadbill (337), of the family of broadbills (Eurylaimidae), widespread in the tropical forests of the Old World, is distinguished by a wide-gaped, flattened bill like all fourteen species of this family. Such a bill is characteristic of birds that capture insects on the wing. The Lesser Green Broadbill also feeds on fruit. It is found in southern Asia.

Calyptomena viridis

The suborder Clamatores which is widely distributed in Central and South America contains the family of cotingas (Cotingidae), numbering

337

338

339

340

ninety-five species of attractively coloured birds that differ in coloration, appearance and way of life. Two of the most handsome birds of the family are the cocks-of-the-rock, noted for their fascinating courtship dance.

The Guianan or **Common Cock-of-the-rock** (338), with narrow black border on the ornamental crest extending from the top of the head to the tip of the bill, is at home in Venezuela, Guyana and the border areas of northern Brazil. The female, with her brownish plumage, is not nearly as handsome as the male.
Rupicola rupicola

The Purple-throated Fruit Crow (339) is one of the lesser known cotingas. It has an ornamental patch of maroon feathers on its throat which it expands during the courtship display. It is found in Costa Rica, Panama and northern South America.
Querula purpurata

The Andean Cock-of-the-rock (340) has deep scarlet plumage but lacks the fine, veil-like feathers on the back. It is found in the northern Andes from Colombia to Peru.
Rupicola peruviana

Both cocks-of-the-rock are typical birds of mountain forests interspersed with rocks. They are fond of localities close to water. The males perform their communal courtship dance on rock ledges; with ruffled feathers, drooping wings and bills practically touching the ground they execute all sorts of hopping antics to the accompaniment of loud cries. The nests of mud lined with leaves and moss are fastened to the walls of caves or placed on rock ledges or in crevices. Frequently the birds nest in small colonies. The female lays two whitish, dark-spotted eggs which she incubates by herself. The male, however, assists in feeding the young.

341

The Red-capped Manakin (341) is one of some sixty species of small, brightly coloured birds
Pipra mentalis
of the family Pipridae, typical inhabitants of South America. Manakins are noted for their courtship display. The male of the illustrated species selects a bare horizontal branch on which to perform his intricate courtship antics.

342

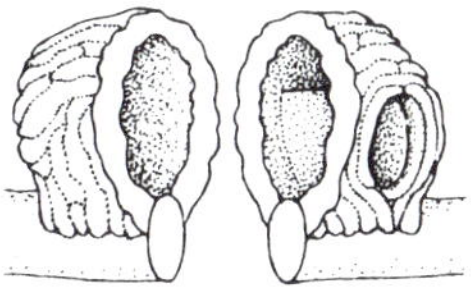

Cross-section of the Rufous Ovenbird's nest

343

The Rufous Ovenbird (342), a drab, brownish bird of the large family of ovenbirds (Furnariidae), inhabits open country in southern Brazil, Paraguay and Argentina. It is known primarily for its skill as a nest-builder. During the rainy season the two partners build the fairly large nest, which resembles a small Dutch oven, on thick branches, fence-posts or the eaves of houses. Made of mud, reinforced with dry plant stems, it is divided into two sections – an entrance-way and nesting chamber, and may weigh as much as ten kilograms. The clutch comprises three to five eggs. The diet consists mainly of insects.

Furnarius rufus

The well-known lyrebirds of Australia (family Menuridae) already exhibit a close relationship to the true songbirds (Oscines) by their vocal abilities as well as by the arrangement of the syrinx. Superficially the two known species of lyrebirds are most like the gallinaceous birds. They are about the size of a pheasant, have strong legs and long toes with long, straight claws, short wings and a bill that also resembles that of the fowl-like birds. They are likewise polygamous and roost in the branches of trees at night, though they spend most of the daylight hours foraging for food on the ground. The young are nidicolous.

The Superb Lyrebird (343, 344) inhabits the thick coastal and mountain forests of southeastern Australia. The note of this grey and brown bird is a simple two to three syllable call but it is an excellent mimic of other bird songs and sounds such as the hooting of a train or car horn, which it intersperses between its own notes. Knowledge of these imitated sounds is apparently inherited for a dog's bark may often be heard in a lyrebird's song even in places where there are no longer any human habitations or never were.

Menura novaehollandiae

The male's most striking feature is the magnificent lyre-shaped tail. It is composed of sixteen retrices, of which the two exterior ones are curved in a V like a lyre, their outside webs narrow, the inner webs broad and ornamented with dark patches. The two central feathers are still and wire-like and without a web, while the remaining feathers, lacking barbules, have loosely hanging vanes with a lacy texture. The

Range of the Superb Lyrebird

females do not have ornamental tail feathers. The striking lyre-shaped tail comes into full play for a period of three to four weeks during the courting season in autumn and winter. At this time the woods echo with the resounding voices of the males as they execute their striking displays in the forest clearings. Before launching the performance the males stake out their nesting territory in which they prepare several display grounds, small spaces about one metre square on top of a mound which the bird has scraped clear of all twigs and leaves. The display begins with the male singing his song perched on the branch of a tree or on a rock; after a while he descends to the ground, continuing his song and performing his dance in the presence of the hens (sometimes even if no hens are present). He spreads his long (up to 75 centimetres) lyre-shaped tail wide and at the climax of the performance tips it forward over his back so that it almost covers him completely. It is only in this position that the silver colouring of the underside of the tail, which otherwise escapes notice, shimmers in all its beauty. In the presence of the female the male produces a drumming sound with the feathers, thus further highlighting the effect of the performance.

The females begin building the nest while the males are still courting. These are large, spherical structures of twigs and leaves, usually placed on or near the ground in the forks of branches, amidst a tangle of tree roots as well as in rock crevices. The side entrance opens in a spacious chamber lined with moss, leaves, grass and feathers in which the female lays the single, large, dark egg. Incubation takes a full six weeks. When

344

345

the chick hatches the female feeds it another six weeks and continues to care for it a long time after it has fledged.

A plump body, large head, short neck, short wings, long, strong legs and abnormally short tail as well as gay, variegated colouring are the distinguishing features of all twenty-three species of pittas (family Pittidae), which inhabit the jungle and tropical forests of southeast and

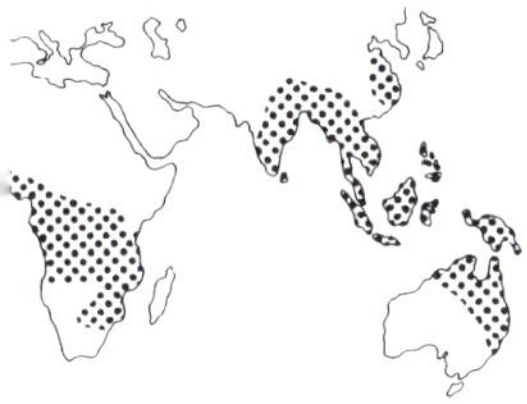

Range of pittas

346

southern Asia and Australia. Two species occur in Africa. Like forest sprites they run about on the ground in the dim light of the jungles, resembling round bundles on long legs. When forced to take to the air they fly only short distances. The nest is globular with a side entrance and is made of twigs, leaves, grass and moss. It is placed amidst a tangle of branches and roots either on or near the ground. Only a few species place the nest high up above the ground. The clutch consists of two to six eggs.

The Banded or **Blue-tailed Pitta** (345) with its variegated plumage is one of the loveliest specimens in the bird world. It occurs in many different races in deep woods, even high in the mountains, from the Straits of Malacca to Java, Bali and Kalimantan. It does not lead a concealed way of life like other pittas, from which it differs in coloration and greater length of the tail. It shows a preference for the tops of rocks and boulders which are a good vantage point for spotting prey.

Pitta guajana

The Blue-winged or **Fairy Pitta** (346) has a wide range extending from Japan across southern China as far as northern India and south as far as the Philippines, where it occurs in several geographical races. It feeds on insects and other invertebrates, for which it rummages with its feet and bill in the fallen leaves and forest litter. The large globular nest (measuring 15 to 26 centimetres in diameter) is built in rock crevices or in the forks of strong branches two to eight metres above the ground. It is made mostly of moss, covered with dry twigs. When disturbed the female conceals the entrance hole with green leaves and tips of branches. The four to six, light-coloured, finely-spotted eggs are incubated by both partners, who likewise share the duties of feeding and caring for the young.

Pitta brachyura

Chapter 28 THE LARGE FAMILY OF SONGBIRDS

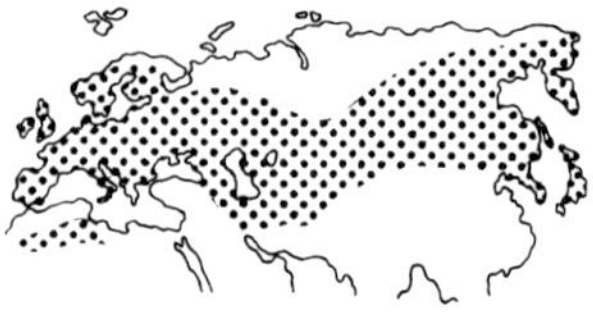

Range of distribution of the Skylark

Range of distribution of the Woodlark

The last group in the system of bird classification is the suborder of true songbirds (Oscines). This large group contains some four thousand species of birds, largest of which is the Raven. Though generally inconspicuous creatures their song has a sophisticated quality and some are among the best singers in the avian kingdom. Others, however, have voices that one would be hard put to call songs. Their young, like those of all the other members of the order Passeriformes, are nidicolous and are dependent on the care of the adult birds for a long time. The nestlings demand food of their parents by opening their beaks wide, thus exposing their brightly coloured gapes, and by uttering high-pitched cheeps. While they are still blind this behaviour is prompted by a jolting of the nest or some movement in its vicinity or else by a change in the intensity of the light on the arrival of one of the parents; later, when their eyes are no longer closed, by the sight of the adult bird. This begging on the part of the nestlings is a necessary impulse that triggers the feeding instinct in the parents. This large group of birds, some species of which are numerically the most widespread, has a cosmopolitan distribution and its members are to be found in all parts of the world.

The family of larks (Alaudidae) includes seventy-five different species, all of which are dull-coloured birds distinguished by having the front and back of the tarsus scaled and a long straight claw on the hind toe. They build a simple, open (very occasionally covered) nest on the ground, gather animal as well as plant food on the ground, and clean their feathers by 'bathing' in the dust or sand. Building the nest and incubating the eggs are tasks performed by the female alone. The eggs are light-coloured and thickly covered with dark spots. Larks are found only in the Old World except for a single species which is native to America.

The Skylark (347), originally an inhabitant of open steppes and mountain meadows above the
Alauda arvensis
forest belt, has over the centuries become a typical inhabitant of fields and meadows. It is one of the few species of birds that brighten the lonely grain fields with their glorious song bubbling from their throats high up in the sky. It is surprising that in our day and age these small songsters are still being caught and shot in large numbers for culinary purposes on their flights to and from their winter quarters.

347

348

349

The Woodlark (348) is distinguished from the Skylark by its shorter tail and pale eye-stripe.
Lullula arborea
It is a typical inhabitant of dry, lowland woods and the margins of mountain forests. Its beautiful song is composed of melodious, melancholy notes interspersed with trills. It may be heard also on clear nights, when it sounds even more beautiful.

The Crested Lark (349) is an inconspicuous bird coloured an earthy brown with erectile crest.
Galerida cristata
Originally a bird of the open steppe and semi-desert it became adapted to life in the vicinity of human settlements, where it runs about rapidly with small steps, occasionally uttering its simple call or delivering its brief song. It is found in Europe, Asia and the northern half of Africa.

350

351

352

Throughout the world there are some seventy-five species of swallows and martins that comprise the family Hirundinidae. They are typical aerial hunters of insects. Though accomplished fliers they walk awkwardly and with difficulty on their short legs.

The Purple Martin (350) inhabits the temperate regions of North and Central America. The sexes have differently coloured plumage. The male is glossy blue all over; the female has pale grey underparts. Even before the arrival of the white man, the Indians helped this fairly large martin by hanging up hollowed gourds for it to nest in; nestboxes, of course, are better. Otherwise it nests in tree cavities. It does not make nests of mud like the Old World martins.

Progne subis

The Swallow (351) has a deeply forked tail, whereby it is easily distinguished in flight from the House Martin, which has a shallowly forked tail and white rump. It is found in Europe, Asia, northwestern Africa and North America. Throughout its range it has become adapted to man's presence and generally builds its nest in a barn, corridor and even in a room. It is an open, cup-shaped structure, constructed chiefly of mud mixed with plant parts. Swallows are migratory and when the second brood has fledged they form large flocks, perch on telephone wires and before their departure for the winter often roost in very large aggregations in reed beds. The eggs (four or five in each clutch) are incubated predominantly by the female for fourteen to sixteen days; both parents, however, share the duties of feeding the young.

Hirundo rustica

The Sand Martin (352) is a small, unobtrusive bird coloured brown and white with slightly forked tail. Unlike the Swallow it builds a simple nest lined with feathers at the end of a burrow which it digs with its bill and feet in sand and mud banks, often near water. The average length of the tunnel, which both partners dig, is sixty centimetres. The female lays five to six white eggs twice a year which she and her mate take turns incubating for twelve to sixteen days. Sand Martins are gregarious birds and nest in colonies which sometimes number several thousand pairs. The Sand Martin breeds in Eurasia and North America and is a migrant, like the Swallow.

Riparia riparia

354

European populations winter in East Africa and Asian populations in the southern parts of Asia and in Malaysia.

A distinctive group of songsters is the family of shrikes (Laniidae), numbering about seventy species. In proportion to their size they are definitely the best hunters among the passerines. This is testified to by their strong, hooked beak which has a notch in the lower bill and a 'tooth' on the upper bill that aids in tearing prey. Shrikes are large-headed and long-tailed birds of practically worldwide distribution, being absent only in South America, Australia and Madagascar. Africa is the home of the greatest number of species. They feed on small animals and smaller vertebrates. Nearly all species have the curious habit of impaling their prey on thorns or hanging their catch in the forks of branches. They build a firm cup-shaped nest in the branches of a tree or bush and both partners share the duties of incubating and rearing the young.

The Red-backed Shrike *Lanius collurio* (353), one of the smaller species, is a typical inhabitant of bushes and forest margins in Europe and Asia. The nest is usually well concealed in a bushy growth or the crown of a small tree. The male and female differ in coloration. European birds migrate in winter as far as South Africa.

The Great Grey Shrike *Lanius excubitor* (354) occurs in many geographical races in North Africa, Europe and Asia. It is also found in northern North America. It generally spends the winter in its breeding grounds but northern populations fly far south. Due to its robust size and skill it can easily capture not only small rodents but even birds. It flies with an undulating flight over its hunting ground, often hovering momentarily in a single spot. The nest is generally located in the top of a tree.

The Loggerhead Shrike *Lanius ludovicianus* (355) is another North American species. It resembles the Lesser Grey Shrike, which also breeds in North America, but is smaller and has a far greater range, extending from southern Canada to southern Mexico and Florida, where nine local subspecies are found. Its life-habits are no different from those of other shrikes. Like them, it often kills more prey than it can consume at a time — an uncommon trait amongst birds — and likewise establishes caches of food. The mainstay of the diet are large insects — grasshoppers, beetles, etc., but it also hunts small lizards, snakes and small mammals. Undigested parts, such as the chitinous armour of insects, bones and scales, are regurgitated. It usually watches out for prey from an elevated spot, sometimes also in hovering flight.

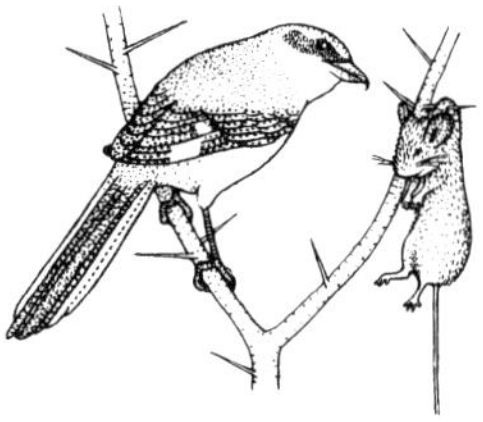

The Great Grey Shrike with impaled prey

355

6 357

358

Of the small family of waxwings (Bombycillidae) comprising eight species, only one is described here.

The Waxwing (356) is a typical bird of the northern forest-tundra and taiga. It breeds in a broad belt extending from Lapland to Kamchatka and in northwestern America. The nest of lichens, moss and grasses is located in the top of conifers. The female is fed by her mate on the nest while she incubates the three or four blue-grey, spotted eggs. When the young have hatched, which takes twelve to fifteen days, they are tended by both parents. In summer the Waxwing feeds mostly on insects, which it captures like the fly-catcher; for the remaining seasons of the year it eats strawberries and other berries. It is a voracious feeder. The vegetable food passes through the digestive tract very rapidly so that often its excreta contain berries and fruits that have been only partly digested. That is why the Waxwing daily consumes twice its body weight in food. In favourable years wax-wings multiply in great numbers and when the ensuing winter is a severe one with a lot of snow immense flocks of these birds fly south as far as southern Europe. Ringing has shown that they fly distances of up to 5,700 kilometres.

Bombycilla garrulus

Closely related to the crows and jays is the family of butcherbirds and bellmagpies (Cracticidae) of Australia. They are very loud-voiced birds which sing in chorus in the morning and evening. Accomplished mimics, they intersperse their own song with imitations of other sounds.

The Pied Currawong or **Crow-shrike** (357) breeds in the dry woodlands of Australia, Tasmania and Lord Howe Island. It feeds on vegetable and animal food.

Strepera graculina

The Common Piping Crow (358) is the most familiar species of this family, its range extending from southern Guinea to central Australia. In places where it is not persecuted it even occurs in the vicinity of human habitations. It feeds chiefly on small invertebrates but also captures various small vertebrates – reptiles, small mammals and birds. It will eat even vegetable food. The nest of twigs and stems is built high in the treetops. The three to four eggs are incubated by the female alone but the young are fed by both parents. The Common Piping Crow lives to an age of as much as twenty years.

Gymnorhina tibicen

359

360

The family of crows and jays (Corvidae) includes both small and large birds; one of them, the Raven, weighing about 1500 grams, is the largest of all living songbirds. The corvids are very intelligent birds with considerable learning and memorizing ability. The reason behind these high mental qualities is their comparatively large brain capacity. Some corvids can even be taught various words. Corvids eat a wide range of food and exhibit marked inventiveness in obtaining it. Comprising some hundred species, they are to be found in practically all parts of the world.

The Jackdaw (359), a nondescript, grey-black bird with peculiar grey eyes, is noted for its cleverness and social behaviour. It is one of the few corvids to nest in cavities and semi-cavities – on cliffs, in hollow trees, in ruins and even in old buildings and chimneys. It nests in large colonies. The four to six eggs are incubated mostly by the female but the young are fed by both parents. During the breeding season the Jackdaw eats mainly animal food, at other times of the year also a great quantity of vegetable food. It nests in Europe and Asia. Northern populations are migratory.
Corvus monedula

The Jay (360) is a familiar bird of the woods in Europe and Asia but is absent in southern Asia. Its harsh cry of alarm warns the other denizens of the woods of approaching danger. It feeds on vegetable as well as animal matter and because, like many corvids, it makes stores of seeds and fruit it contributes towards the spread of certain plants and trees. The cup-shaped nest is located in trees and bushes and the clutch consists of five to seven eggs.
Garrulus glandarius

The Nutcracker (361) is found in the coniferous forests of Europe and Asia. Some years, northern and Siberian birds invade central and western Europe in large numbers, apparently for lack of food, filling the woods with their harsh cries, resembling those of the Jay.
Nucifraga caryocatactes

The Blue Jay (362) is the best known North American jay because of its bright colouring as well as its intelligence. It lives in woodlands east of the Rockies but may also be seen in parks and on the outskirts of many cities.
Cyanocitta cristata

361

362

363

364

The Red-billed Magpie (363), like many corvids, is a good mimic of other birds and with its harsh cry of alarm warns other birds of approaching danger. Its own song, heard during the courting period, is soft and pleasant. It inhabits the woodlands of southeastern Asia where it feeds on vegetable matter and small animals, often stealing eggs from the nests of other birds.
Urocissa erythrorhyncha

The Steller's Jay (364) is one of the most beautifully coloured birds of western North and Central America. It inhabits woods, moving, in winter, close to human habitations. The diet consists of insects, small invertebrates and vertebrates, birds' eggs, fruit and seeds. It is a noisy, agile bird.
Cyanocitta stelleri

The Rook (365), a black bird with blue metallic sheen and bare area round the base of the bill, is often the only inhabitant of the barren winter landscape. It always occurs in large flocks, also nesting in large colonies, often numbering thousands of pairs. It is true that rooks cause damage to fields of sprouting grain, but on the other hand they destroy a great many harmful insects and other small animals as well as rodents. Their usefulness definitely outweighs any damage they may cause. The Rook breeds from Central Europe through all of Siberia to the Far East.
Corvus frugilegus

365

366

The Hooded Crow (366) is just a geographical race of the all-black Carrion Crow *(Corvus corone)* widespread through the whole Palearctic region, with grey and black plumage. It breeds in the region between the River Elbe, Bohemia and the Yenisei; the area west and east of this territory is the home of the Carrion Crow. The Hooded and Carrion Crow often interbreed. The Hooded Crow feeds on animal and plant matter and causes much damage to the clutches of other birds.

Corvus corone cornix

The Raven (367) nests on cliffs and in tall trees and feeds mainly on carrion and refuse of various kinds. Its range originally included all of Europe, North and East Africa, practically all of Asia and North America; in Europe, however, it has been exterminated in many areas.

Corvus corax

367

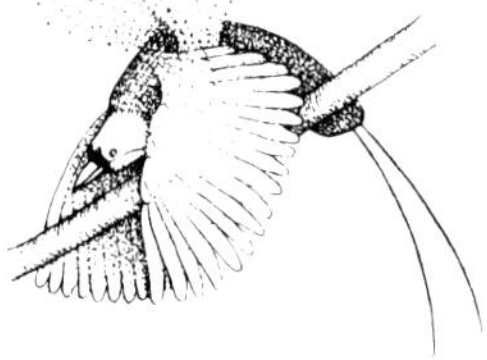

Courtship display of the Red-plumed Bird of Paradise *(Paradisaea apoda)*

368

Most of the forty known species of birds of paradise (Paradisaeidae) have gorgeous and colourful plumage but their song is not one of their strong points. The great demand for their magnificent plumes proved a disaster for some species. Today, however, these birds are protected by law in their native territories — New Guinea and certain adjacent

369

islands, the Moluccas and northeastern Australia. Birds of paradise spend most of their time in the branches of trees, the males' courtship display also being performed there. During this performance they sometimes adopt fantastic poses, spreading or raising their plumes and uttering frequent loud cries as they parade before the females. Birds of paradise build their cup-shaped nest in trees and usually lay two eggs. During the breeding period the males devote all their time to showing off, leaving the task of incubating and rearing the young to their soberly-clad mates.

The Wilson's or **Waigeo Bird of Paradise** (368) boasts six colours and various ornaments, the most distinctive being the crossed and curving wire-like central tail feathers. This small bird of paradise lives in the forests of New Guinea and certain outlying islands. The male performs his colourful courtship antics on an upright trunk or branch growing out of the ground as well as on the ground itself. Round the base of the branch he clears several metres of ground of all leaves and twigs and also nips the leaves off the 'display' branch.

Diphyllodes respublica

The Red Bird of Paradise (369) is one of the large species in which the males have flag-like plumes on the flanks and long, wire-like tail feathers curving in a spiral. It is found only on the islands of Waigeo and Batante off the west coast of New Guinea.

Paradisaea rubra

370

Some species of the family of bowerbirds (Ptilonorhynchidae) which breed in the forests of Australia and New Guinea, build, as part of their courtship, an elaborate structure in and near which the birds dance. The bowerbirds are closely related to the birds of paradise and, like them, feed on fruit, seeds and small invertebrates. Though they are good fliers they spend most of their time on the ground. The cup-shaped nest is built in a tree fork and the clutch comprises two or three eggs. The female incubates and rears the young by herself.

The Great or **Red-crested Bowerbird** (370) is about the size of a Jackdaw and lives in the bush
Chlamydera nuchalis
country of northern Australia. The male and female have like plumage; the male usually has elongated purplish feathers on the nape, which when expanded during the courtship display look like the flower of a cactus. The male spends a full two months before the courtship preparing the display ground. Selecting a spot under a suitable bush he carefully removes all leaves, twigs and stones and then begins building an odd structure resembling a bower. First he prepares a layer of twigs about ten centimetres high and into either end of this platform he inserts upright twigs of about the same length. The resulting structure is forty to fifty centimetres long and forty-five to fifty-five centimetres high with an inside tunnel about twenty-five centimetres high and fifteen centimetres wide. The space in front of each of the tunnel openings is further decorated with all sorts of objects such as shells, bones, stones, bits of

371

372

glass, metal and, in places close to frequented paths, also bottle caps. Always, however, the ornaments are coloured either white or grey. When the structure is completed, the male begins singing from his perch on a nearby branch. As soon as he succeeds in attracting a female he begins the actual courtship display inside the bower, ruffling his feathers, bowing, expanding the feathers on his nape and uttering various cries. All this leads to the act of copulation, which takes place inside the tunnel.

The Satin Bowerbird *Ptilonorhynchus violaceus* (371) measures about thirty centimetres and is at home in the damp forests of eastern Australia. Its black feathers have a violet-blue metallic sheen, and this is probably the reason why the bird selects blue-coloured objects to decorate its bower. Its instinctive behaviour is even more complex than that of the Great Bowerbird for it daubs the inside walls of the bower with a bluish paint made of crushed blue berries and charcoal mixed with saliva.

The most familiar local songbirds in Africa, Madagascar and all of southern Asia are the bulbuls (Pycnonotidae). Of the 112 species best known are those that have adapted themselves to life in the vicinity of human habitations. Bulbuls are small but active and conspicuous birds sombrely clad in grey, brown or olive green. They feed on fruit and insects. The somewhat haphazard nest of twigs and roots is built in the branches of a tree or bush and the clutch comprises two or three eggs.

3

37

The White-cheeked Bulbul (372) is a common, cheerful and trusting bird found near human settlements from Mesopotamia to India. It is fond of perching in an elevated spot where it utters its cries or watches for insects. The picture shows it in such a typical pose.

Pycnonotus leucogenys

The Red-whiskered Bulbul (373) is one of the more brightly-coloured species. It is a native of India, China and southeastern Asia and likewise a typical inhabitant of villages and towns. It nests not only in trees and bushes but also in climbing plants on houses. The song is composed of pleasant flute-like notes and less pleasant cries.

Pycnonotus jocosus

The Fairy Bluebird (374) shown in the picture is a female. The male, with his pale blue and velvety black plumage, is even more beautiful. It lives in the damp forests of southeastern Asia and India and builds its nest in a bush. It feeds solely on strawberries and other fruit and is fond of collecting the nectar from flowers. When doing so pollen often clings to the feathers under its beak, thereby being transferred to other flowers.

Irena puella

The Black Drongo or **King Crow** (375) is a solitary, glossy black bird measuring twenty-seven centimetres, which makes its home in the forests of India and China. It is very adroit at catching insects on the wing. In its native habitat it often perches in an elevated spot or on the backs of cattle to catch insects which they stir up. The Black Drongo is a member of the family Dicruridae, widespread in Africa and southern Asia to Australia. It contains twenty-one species of birds, all of which have forked tails.

Dicrurus macrocercus

375

376

377

The Bald Crow or **Bare-headed Rock Fowl** (376), about which little was known until recently, is a bird of the virgin forests of western Africa and is so called because its head is almost entirely bald. It inhabits rocky country where it hops about on its long legs, scattering the leaves with sharp movements in its search for arthropods and small vertebrates. The cup-shaped nest,

Picathartes gymnocephalus

378

Range of distribution of the Bald Crow

379

made of hardened mud, is located in semi-cavities and rock crevices. The two young are fed regurgitated food. The Bald Crow is classed in the family Timaliidae which also includes the following species.

The White-crested Laughing Thrush (377), measuring about twenty-eight centimetres and possessing an erectile crest, breeds from the Himalayas to southwestern China and Cambodia and in western Sumatra. The shallow nest of roots and grasses is built near the ground in bushes. The four or five eggs are incubated by both parents.
Garrulax leucolophus

The Bearded Tit (378) (the picture shows young nestlings) is a cinnamon coloured bird with long tail. Originally classed with the Timaliidae it is now placed in a separate family by itself (Paradoxornithidae). It inhabits extensive reed beds where it conceals its ragged nest, placed close above the ground or water. In the dim light of this environment the young birds' red gapes edged with yellow and marked with white spots serve as a visual stimulus that triggers the feeding instinct in the adults. The Bearded Tit has a discontinuous distribution reaching from central Europe through central Asia to the Amur region.
Panurus biarmicus

The Reedbed Parrotbill (379) was known to breed only in a very circumscribed area on the middle reaches of the Yangtse river until recently when another breeding ground was discovered in the Far East by Lake Khanka north of Vladivostok, where this picture was taken. The Reedbed Parrotbill feeds largely on insects which it collects from aquatic plants or nips out of reed stems with its strong, laterally flattened beak used like scissors.
Paradoxornis heudei

380

The Pekin Robin (380) has a lovely, melodious song and attractive coloration. It is found in jungles and forests at higher elevations from the Himalayas to south-eastern China. The small insects and other arthropods as well as straw-berries and seeds which it eats are gathered both in trees and on the ground. The sturdy nest of moss, lichens and leaves is placed in tree branches as high as three metres above the ground. Both parents take turns incubating the three or four eggs for twelve davs. The young leave the nest at the age of eight to ten days.

Leiothrix lutea

381

The Red-breasted Flycatcher (381) belongs to the large family of flycatchers (Muscicapidae) which numbers some 330 species. The reddish-brown breast and large eyes are reminiscent of the Robin. Unlike the latter, however, the Red-breasted Flycatcher stays high up in the tops of trees and only its two-syllable call or brief song reveals its presence. The nest is located in holes or semi-cavities in trees high above the ground. The breeding range extends from central Europe eastward as far as Kamchatka and Sakhalin. The wintering grounds are located in southern and southeastern Asia.

Ficedula parva

382

383

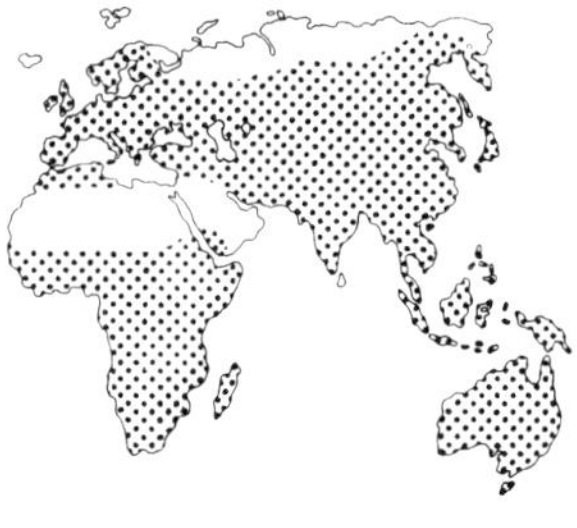

Range of Flycatchers

Flycatchers (Muscicapidae) generally hunt flying insects. Perched upright in an elevated spot they watch for their prey, darting out in pursuit as soon as they spy it and often returning to the same perch. The broad, flat bill is well adapted to this method of hunting, and the short, tough bristles at its broad base are a further help. The finely structured legs and feet are suited only for perching on branches. Flycatchers are widespread in Eurasia, Africa, Oceania and Australia.

The Spotted Flycatcher (382), an inconspicuous, sombrely clad grey bird, is common in mixed and deciduous woodlands, parks, gardens and tree groves. It nests in semi-cavities. The nest is a solid structure of blades and roots lined with hairs and feathers. The clutch usually consists of five greenish, thickly spotted eggs and the young hatch after thirteen days. Even the layman recognizes it as a flycatcher by its method of hunting and the way it flicks its wings and tail as it alights. One only hears its soft song rarely, but when hunting it frequently utters a number of various calls. It is at home not only in the forests of Europe and Siberia but also in North Africa. It winters in tropical and southern Africa.

Muscicapa striata

The Collared Flycatcher (383) differs from the preceding species in having black and white plumage. It inhabits leafy groves, where it nests in tree holes and man-made nest boxes. It is an excellent agent in the control of harmful insects. It has a discontinuous distribution in western and southern Europe and occurs in a continuous range from central Europe to the middle reaches of the Volga River.

Ficedula albicollis

The Narcissus Flycatcher (384) is a typical bird of the coniferous and deciduous subtropical forests of eastern Asia and Japan. A migrant, it winters in Malaysia and on the Sunda Islands. As in the Collared and other flycatchers there is a marked difference between the sexes, the striking yellow colour of the male being completely absent in the female.

Ficedula narcissina

384

385

386

Many warblers of the family Sylviidae may be said to be master singers. However, this great talent is not matched by their coloration. They are small, largely inconspicuous birds. The predominant colour is green. The fine, awl-shaped bill is slightly downcurved at the tip. The diet consists chiefly of insects, though the birds feed on other invertebrates as well. They are tireless and active birds that are continually moving through the thick branches of trees, bushes, grasses or reeds. Most of the 400 species are to be found in the Old World; only a small number occur in the New World.

The Long-tailed Tailorbird (385) is noteworthy for the way it prepares its nest. It bends larger green leaves or several smaller leaves into a sort of cup, then pierces holes in the edges with its beak and sews the leaves together by drawing threads (plant fibres, silk from spiders' webs, etc.) through the holes. It then builds its nest of plant or animal down inside this cup, in which the female lays three eggs. This method of sewing nests is characteristic of several species of warblers but the Long-tailed Tailorbird exhibits great proficiency. During the breeding period the two central tail feathers are greatly elongated in the male. The Long-tailed Tailorbird breeds in India, southern China to Malaysia and in Java. It may be found even in the vicinity of human habitations.

Orthotomus sutorius

The Red-headed Fantail Warbler (386) builds its spherical nest in a thick tangle of herbaceous plants and grasses, sometimes 'sewing' it to the plant leaves. The entrance is at the side. The bird's breeding plumage has shorter tail feathers than the winter garb. The Red-headed Fantail Warbler breeds from India and southern China as far as eastern Australia.

Cisticola exilis

387

Warblers of the genus *Sylvia* number nineteen species that are at home in Europe, southeast Asia and some even in North Africa. They are insect-eaters but in autumn also eat berries and fruit in large quantities. Some species even feed fruit to their young when insects are scarce.

The Blackcap (387) is confined to woods more than any other European warbler. Its song, starting with a grating note and changing to a mellow, flute-like warble of rising strength, is usually delivered from treetops. The nest, however, is placed in thickets near the ground. The male has a small black cap, the female a reddish brown cap. The picture shows the Blackcap removing its offspring's droppings. The Blackcap breeds in Europe, western Siberia, the Middle East and northwestern Africa.
Sylvia atricapilla

The Barred Warbler (388) shows a great fondness for thorny thickets in which it conceals its simple but sturdy nest. Its song is rapid and harsh, interspersed with its call note and now and then with other bird songs which it imitates. It breeds in Europe and southwestern Asia.
Sylvia nisoria

The Whitethroat (389) is a sombrely coloured, modest bird that nests in thickets, overgrown ditches, hedgerows, nettles and green hedges. The male usually delivers his rather high-pitched and not too pleasant warble from branches but now and then he does so in song-flight, rising briefly in an arc above the bushes. Both birds build the nest which is located in thick growth near the ground. The Whitethroat breeds in Europe, North Africa and east as far as Lake Baikal. It is a migratory bird.
Sylvia communis

The Bush Warbler (390) is a bird of the Far East where it leads a concealed life in dense thickets near water as well as on mountain slopes. The male stands motionless as he delivers his mellow, flute-like song. The clutch consists of four or five rufous eggs.
Horeites diphone

388

389

391

A concealed way of life is also typical of the grasshopper warblers (genus *Locustella*) which have graduated tails and are more often heard than seen. Their monotonous chirping or rasping songs sound like the notes of an insect.

The Gray's Grasshopper Warbler (391) breeds from the River Ob region eastward to Sakhalin and winters in the Sunda Islands. It works its way through thick growths of grass and bushes. The male begins his song on the ground, climbing, as he sings, to the top of a bush and then leaping down again to the ground. The nest, a fairly large structure with a deep hollow, is placed in dense undergrowth near the ground.
Locustella fasciolata

The Short-tailed Bush Warbler (392) resembles the wren in its size, short tail and movements. It has a fairly small range in eastern Asia and Japan.
Urosphena squameiceps

392

393

The Icterine Warbler (393) inhabits open deciduous woods, parks and gardens from northern France, Switzerland and southern Scandinavia to western Siberia. It winters in tropical and South Africa, arriving at its breeding grounds in early May and leaving again in August. The Icterine Warbler is an excellent singer. The song consists of its own, repeated phrases as well as the songs of other birds living in the neighbourhood. The nest of plant stems and leaves interwoven with cambium, spiders' webs and birch bark, is placed in the fork of a branch in a bush or tree one to three metres above the ground.

Hippolais icterina

394

The Sedge Warbler *Acrocephalus schoenobaenus* (394) belongs to the large group of grass warblers (genus *Acrocephalus*) which are bound to beds of reeds and timothy grass beside water at least during the nesting period. The Sedge Warbler is partial to places where reed beds are adjoined by tall grass interspersed with bushes on which the male perches as he sings his song, composed of various chirping notes, whistles and trills, flying up in the air now and then in the manner of the Whitethroat. The basket-like nest is woven between plant stems close above the water or ground. The Sedge Warbler breeds in Europe eastward as far as western Siberia.

395

396

397

The Marsh Warbler (395), unlike other warblers, often occurs far from water, however, always in wet locations with dense vegetation. For lack of other opportunities it also nests in fields of grain, rape or clover. Its bright song, reminiscent of the Icterine Warbler's, may be heard most of the day and often even at night. It also imitates the notes of other birds. The nest is built in the typical manner of warblers, but always above dry ground. The Marsh Warbler breeds in Europe and western Siberia and winters in East Africa.
Acrocephalus palustris

The Great Reed Warbler (396) inhabits reed beds where it skilfully clambers amongst the reeds, perching on the top of a reed stem only when delivering its song. This is easily distinguished from that of other warblers in that it is loud and strident and made up of various harsh, grating sounds. The basket-like nest is suspended fifty to ninety centimetres above water on reed stems which pierce the walls and serve as a support. The nest itself, up to twenty centimetres high, is made of various dry stems and leaves. In Europe the Great Reed Warbler raises only one brood, the clutch consisting of four to six eggs which both parents incubate for fourteen to fifteen days. When they are twelve days old the young leave the nest though as yet unable to fly. The Great Reed Warbler breeds practically throughout the whole Palearctic region. It winters in tropical Africa and southeast Asia.
Acrocephalus arundinaceus

The Chiffchaff (397) is one of the commonest of birds in European forests. In spring its monotonous song, made up of the continually repeated syllables 'zilp zalp', may be heard in the lowlands as well as hills. It breeds in Europe and Siberia, the Middle East and northwestern Africa.
Phylloscopus collybita

The Willow Warbler (398) is a typical bird of the forests of Europe, northern Asia and areas far to the north. It is best distinguished from the Chiffchaff by its song, which is a clear and descending warble totally unlike the Chiffchaff's.
Phylloscopus trochilus

All willow warblers (genus *Phylloscopus*), which include some fifty species, build spherical nests with entrances at the side placed in a tangle of vegetation either near or on the ground. They are built mostly by the females.

398

399

Eurasia and North America are the home of five species of kinglets (genus *Regulus*). Their weight of five grams makes them the smallest of Eurasian birds. Very active, they are continually fluttering among the treetops in pursuit of insects and spiders, sometimes hovering briefly in a single spot. They are not fond of traversing greater distances. The fairly large, warm nest of moss, lichens and spiders' webs is located in dense branches.

The Firecrest (399) has distinctive markings on the head and is more brightly coloured than the Goldcrest. It is partial to mixed and deciduous woods and is found in Europe, Asia Minor and North America. The seven to eleven eggs are incubated by the female alone, but the young are tended by both parents.
Regulus ignicapillus

The Goldcrest (400) is one of the commonest birds of the forests of Europe and Siberia, even though it often escapes notice because of its concealed way of life in the
Regulus regulus

400

401

treetops. It remains in the woods even in severe winters, searching for food together with flocks of tits.

The Robin (401)
Erithacus rubecula

In the dusk of a spring evening the Robin's melancholy, warbling notes may be heard together with those of Song Thrushes, Mistle Thrushes and Blackbirds almost until nightfall. All these songsters, together with some three hundred other species, belong to the family Turdidae, found practically the world over. The male Robin always delivers his song perched high up on a branch, though Robins spend most of their time on the ground where they seek the animal food they feed on in the shadow of the undergrowth. They build their large sturdy nest in fallen trees, tree stumps or other cavities and have two or three broods a year. Both partners tend the young. The Robin breeds in woodlands from Europe to western Siberia, its range extending as far as North Africa and Asia Minor but it is also a familiar garden bird in the British Isles. Northern populations are migratory whereas south and west European birds stay the winter.

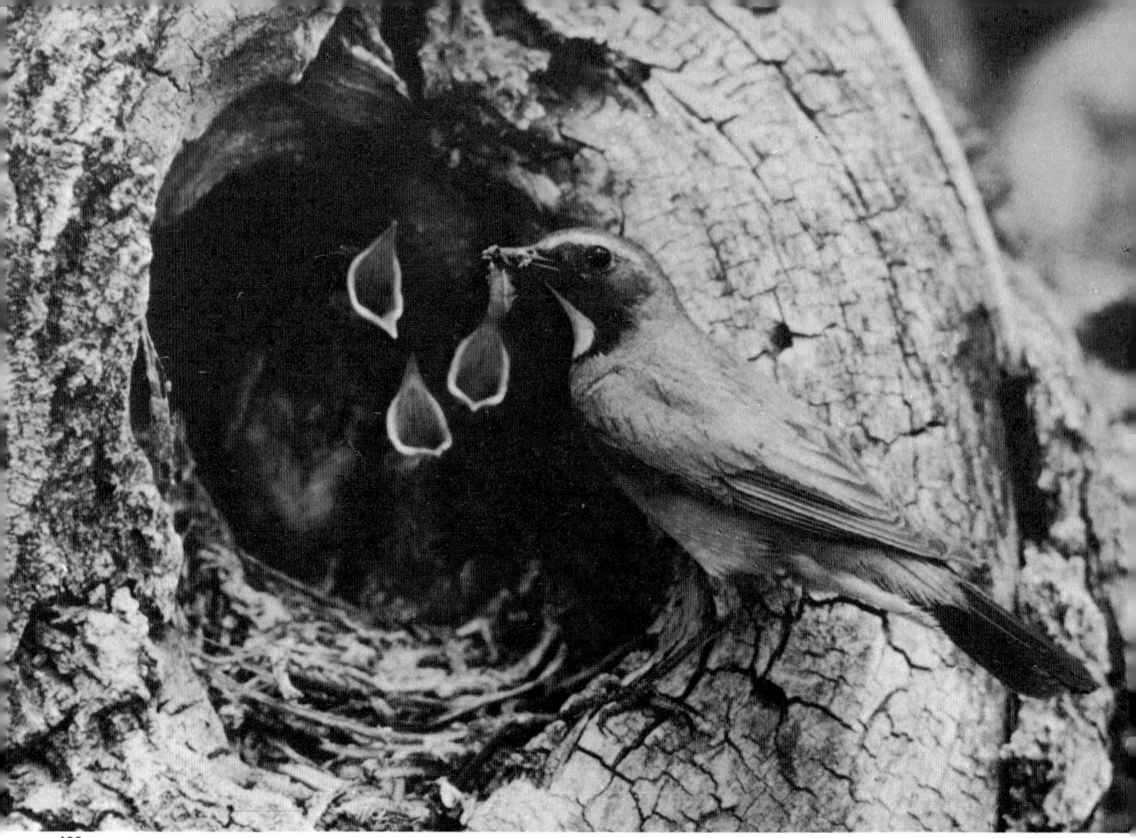

402

403

404

The White-throated Robin (402) has a fairly small range, being found in the Middle East, Iran, part of central Asia and Afghanistan. It is a shy bird that makes its home in thickets on stony mountain slopes and at the edges of dry woodlands. It moves through the thickets in the manner of warblers, but the flicking of its tail and wings is reminiscent of the Redstart.
Irania gutturalis

The Wheatear (403) is a typical inhabitant of open country — the steppe, desert and rocky mountain slopes. At lower elevations it nests in stone walls, in quarries and in sand pits. The nest is concealed in various holes and crevices. The Wheatear spends most of its time on the ground, flying from one boulder to another and perching in various elevated spots, staying up in the air for a greater length of time only when performing the courtship flight. The Wheatear is found in Europe and in Asia eastward through the Middle East and central Asia to the Chukotski Peninsula. It breeds also in the western and eastern parts of the extreme north of America. Like the European birds American populations winter in Africa. Alaskan birds fly there across Asia, Greenland birds across the Atlantic and Europe.
Oenanthe oenanthe

The Black Redstart (404), originally a bird of rocky country both in the lowlands and mountains, in time became an associate of man, building its nest on human dwellings instead of rocks. In the mountains it is found even high above the tree line, where its song is often the only sound to break the hushed stillness. Otherwise the Black Redstart is a familiar bird of house gables and chimneys. It nests under eaves as well as in rock crevices. It is found in Europe, the Middle East and central Asia and in northwestern Africa. Northern birds are migratory, wintering in the Mediterranean and North Africa.
Phoenicurus ochruros

405

The Nightingale (405) is rightfully considered the best singer of all European birds. Its joyous, musical song filled with trills and clear ringing notes may be heard both during the day and at night, having a special magic in the hours of darkness. The Nightingale inhabits thickets or the undergrowth of woods where it seeks animal food on the ground like the Robin. Dense undergrowth is also where it builds its well-camouflaged nest of dry leaves. It is erected by the female, who also incubates the five olive-brown eggs by

Luscinia megarhynchos

406

Migration routes of European Nightingales

407

herself for thirteen days. The Nightingale breeds in Europe, southwest Asia and northwest Africa.

The Bluethroat (406) is closely related to the Nightingale even though it is much more attractively coloured. It is found across northern Europe ranging eastward in a broad belt as far as eastern Asia. It occurs in several geographic races. Shown in the picture is the Scandinavian Bluethroat with red patch on the breast. The Bluethroat is fond of nesting on the ground in dense, low vegetation near water.

Luscinia svecica

The Redstart (407) is a more beautiful counterpart of the Black Redstart. It nests in various holes and semi-cavities, mostly in deciduous and mixed lowland woods as well as in coniferous forests up to the tree line. Its range extends from northwestern Africa across the whole of Europe to central Siberia.

Phoenicurus phoenicurus

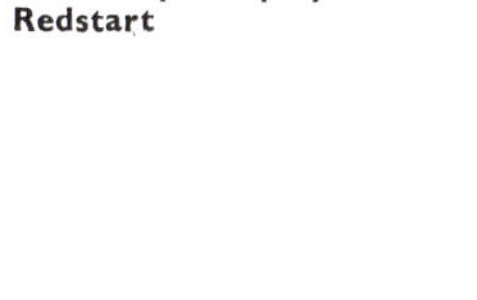

Courtship display of the Redstart

408

409

The White-throated or **Forest Rock Thrush** (408) is as yet a little known inhabitant of the coniferous and mixed woods of the Far East. Unlike the other members of the genus *Monticola* it is not confined to rocky country.
Monticola gularis

The Whinchat (409) is a bird of open country in lowlands, hills and mountains. It is generally met with in wet or swampy meadows with the odd bush, small tree or taller plant amidst the tall grass. It is from such an elevated spot that the male delivers his raucous spring song interspersed with various
Saxicola rubetra

410

whistly notes and imitations of other birds. The nest, built by the female on the ground, is always well concealed in thick vegetation or under a bush. The Whinchat breeds in Europe, its range extending in a narrow belt as far as the Yenisei.

The Eastern Bluebird *Sialia sialis* (410) is a beautifully coloured bird, at least the male is, with his handsome blue mantle, chestnut breast and white underparts. The female is greyish-blue above. In eastern North America and in Central America, home of this elegant bird, it is just as well-known and as popular as the robin is in England. The simple but pleasant warbling song, consisting of three to eight notes, may be heard not only in open woodlands but also in parks near human habitations. Originally a cavity nester it will also raise its young in a man-made nestbox. It has two or three broods a year. The female incubates by herself but the male helps rear the offspring. When they are fully-grown the parents often select different partners for further nesting. They feed on various insects, chiefly grasshoppers and beetles, as well as fruit, especially in the colder months. They catch insects on the wing but not nearly as adroitly as the flycatchers.

412

The Fieldfare (411) is the only thrush that exhibits social behaviour throughout the year. It is a colonial nester and during the breeding season always several nests are placed close together in trees in field groves or the edges of woods surrounded by wet meadows. The song is a grating twittering usually delivered in flight. The Fieldfare is widespread from central Europe and Scandinavia to Lake Baikal and the middle reaches of the River Lena. Large flocks of these birds migrate to their winter quarters in central and southwestern Europe from regions beyond the Urals.

Turdus pilaris

The American Robin (412), the best known and best loved bird of North America, is man's friend from Alaska to southern Mexico. It has grey upper parts, rusty red underparts and conspicuous white spots round the eyes. It is a good singer, tirelessly sounding its musical whistling note. It has two broods a year, the clutch consisting of four to five bright blue eggs.

Turdus migratorius

The Mockingbird (413) belongs to the family Mimidae, found only in America and closely related to the Turdidae. These birds are all good singers and excellent mimics, their song being heard throughout the year the same as the

Mimus polyglottos

413

414

wren's. These uniformly grey birds with white-barred wings and graduated tail are common in the southern United States and Mexico.

The White-rumped Shama (414), with its extremely long, graduated tail, is at home in India and Indonesia. It inhabits the dense jungles where it nests in tree holes or amidst a tangle of roots near the ground. The ground is also where it mostly gathers the various insects it feeds on. An excellent singer, it is often kept as a cage bird in its native land.
Copsychus malabaricus

The Wren (415) is the only member of the large family Troglodytidae (numbering sixty-two species) that lives in Europe, Asia and North Africa; all the others are typical birds of North and South America. It is a small and active, rufous-brown bird with short tail cocked upward and short wings. It moves expertly in its favourite habitat – very thick undergrowth and tree roots in gullies and alongside brooks. It is an extremely inquisitive bird and whenever it sees something suspect it immediately utters its buzzing alarm note 'zerrrr'. Its sharp and simple song is surprisingly loud and is one of the few to be heard even in the middle of winter. In spring the male builds several nests for his mate to choose from, the others serving either as sleeping quarters or as a place of concealment. They are spherical structures made of moss with the entrance hole at the side.
Troglodytes troglodytes

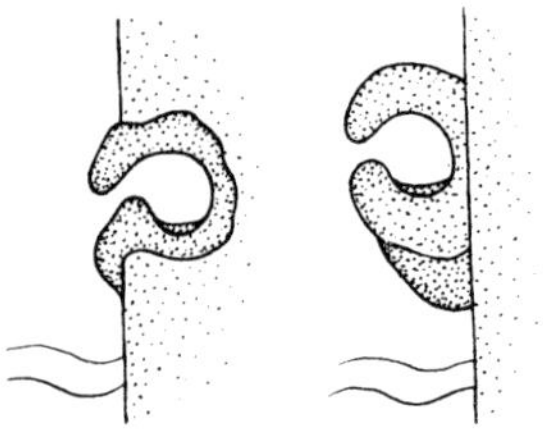

Cross-section of the Dipper's nest

418

The Wren is generally resident, which is perhaps the reason why it forms some forty races throughout its range. More northern populations, however, migrate over a thousand kilometres southward for the winter.

The House Wren (416) *Troglodytes aedon* is one of the commonest of the American wrens, ranging from southern Canada to Mexico and found in mountains and woodlands as well as gardens and near human dwellings. It nests in all sorts of places, often very peculiar ones, but usually in tree cavities and nestboxes. It often has as many as three broods a year. The clutch consists of five to six white eggs finely spotted with reddish-brown which the female incubates by herself. That is why the male is sometimes polygamous.

The Dipper (417) *Cinclus cinclus* together with four other species found in Eurasia, North Africa and western America is the only songbird leading an aquatic life. It and its relatives are found along stony brooks and streams in the mountains and foothills. They are able to run about on the stream bed even in rapidly flowing water and can remain underwater for up to thirty seconds. The nest is a large sturdy structure of moss and aquatic plants placed under a bridge, in the wall of a mill race or on a steep river bank, often even under a waterfall. It has two broods a year, each comprising four to six young that can swim and dive well before they are able to fly. The Dipper stays the winter in its breeding grounds.

419

The Tree Pipit (418) is one of thirty-nine species of brown-spotted birds that spend most of their
Anthus trivialis

time on the ground and belong to the same family as the wagtails. The Tree Pipit is fond of perching in trees but like its relatives nests on the ground. The clutch comprises four to six light brown eggs spotted grey and red; they show such differences in coloration that no two clutches are alike. A characteristic feature of the pipits is the delivery of their song in flight beginning with a melodious twitter and ending with descending peeping notes as they drop to the ground with widespread wings. The Tree Pipit is a migrant and European birds winter in the Mediterranean.

The Grey Wagtail (419), like all wagtails, is distinguished by a slender body and long tail, which
Motacilla cinerea

it pumps up and down for a while after alighting. Unlike pipits, the wagtails' nuptial plumage differs from their non-breeding plumage, the males being more forcefully coloured. Wagtails may be identified by their flight, which is undulating, with the birds uttering their clear metallic call at every swing. The Grey Wagtail has a discontinuous distribution in western and central Europe, in central Asia to Japan and in North Africa. It is fond of fast-flowing streams in lowland country as well as in high mountains. The nest is generally located on rocks beside water, on eroded banks between tree roots, and often on various constructions by the water's edge. The clutch is five or six ochre coloured eggs covered thickly with darker spots. The birds rear two broods a year. Southern populations are resident, northern birds migrate southwest to the Mediterranean and North Africa; individual vagrants may also be met with beside streams in winter in central Europe.

The White Wagtail (420) is the commonest of the wagtails found throughout most of Eurasia, including Japan, and the coastal regions of northwestern Africa, being absent only in the extreme northern arctic. It prefers locations near water but may be encountered also in fields and pastures, in mountains even far above the tree line. It is often found near human habitations. The nest is generally located on various man-made constructions, sheds, foot-bridges, or terraces, but may also be sited in tree holes, rock crevices or even on the ground. Twice a year, in April and in June, the female lays five or six whitish eggs densely covered with grey-brown spots. The White Wagtail is a frequent host of the Cuckoo. In the autumn large flocks gather to roost in reeds, the same as Starlings and Swallows. Northern populations are migratory, their winter quarters being the Mediterranean and North Africa. They are one of the first heralds of spring.

Motacilla alba

420

42

42

423

The Yellow Wagtail (421) has a shorter tail than the Grey Wagtail. In Europe, Asia, Alaska and
Motacilla flava
North Africa it occurs in twenty-two races differing in the head colouring of the male in his breeding plumage. The picture shows the north European form *M. f. thunbergi* with dark head. The Yellow Wagtail is found in fields, meadows and on fallow land; the nest is always placed on the ground. It winters in the Mediterranean and tropical Africa, eastern populations in southern Asia.

The Forest Wagtail (422) is an inhabitant of the oak woods of eastern Asia. It is a sort of tran-
Dendronanthus indicus
sitional form between the wagtails and pipits. Like the pipits it has the same plumage in spring and summer and the sexes are alike in colour. However, it nests in trees, placing the open nest on a horizontal branch.

Titmice are a family that includes several of the commonest of all woodland birds. All have short wings, so that they are not good fliers, and stout, pointed bills useful both for extracting insects from bark and the oily kernels from various seeds in the winter. They utter various notes but cannot be credited with song. They number forty-five species occurring in many geographical forms throughout the tree belt of the Palearctic. For forestry and fruit growers tits are the most natural and best means of controlling harmful insects.

The Crested Tit (423) is a less common tit, confined almost exclusively to coniferous woods. Its
Parus cristatus
presence is revealed by its ringing, warning note – a repeated 'zi-gurr'. It is at home in Europe, its range extending from the Pyrenees and Alps to the Urals and northward to the Arctic Circle. It nests in tree holes, or rotting stumps, and like all tits is fond of man-made nest boxes. The seven to ten eggs are whitish with reddish-brown spots. There are usually two broods a year, the second generally more numerous than the first.

424

425

426

The Blue Tit (424) is the second most common tit next to the Great Tit, the two often being
Parus caeruleus

found together. It prefers deciduous woods and is common in parks and gardens. It avoids coniferous woods, and does not occur high in the mountains. It has a smaller range than the Great Tit; besides Europe it is found in North Africa and the Middle East, reaching as far east as the Urals. It nests in tree holes and generally lays ten to fourteen eggs coloured the same as those of other tits. In winter it often roams the countryside in flocks with other tits. Some central and north European birds, mainly young birds, migrate to Italy, France and Spain.

The Coal Tit (425) is common in all coniferous woods from lowland country to the tree line high
Parus ater

in the mountains. It is widely distributed from the Atlas Mountains in North Africa across all of Eurasia to Kamchatka and Japan. The male's repeated call 'weetse weetse weetse', often changing in rhythm, may be heard short distances apart already in the first days of early spring. The nest is built in tree holes, but because there are not enough natural cavities in today's coniferous woods the Coal Tit often locates its nest in a rotting stump or ground hole. The clutch is eight to ten eggs and the birds raise two, very occasionally three broods a year. In winter they roam the countryside sometimes flying several hundred kilometres, and often occur in large numbers even in city parks and gardens.

The Great Tit (426) is the commonest tit and least particular as to environment, being found
Parus major

in woods as well as in city parks and gardens. It is widespread throughout Eurasia, except the treeless areas, as far as the Sunda Islands and also in northwestern Africa. A characteristic feature of this tit is the penetrating, metallic call of the males in several different rhythms, most familiar being 'ci ci ba' and 'ti ti si'. The nest is located in all sorts of

427

428

tree holes and other cavities. The clutch is 8 – 12 eggs and there are two broods a year. Incubation is performed only by the female but the male assists in feeding the young.

The Black-capped Chickadee (427) is found in many parts of Europe and Asia as well as North America. It is a nimble little bird continually sounding its call note 'chickadee-dee-dee', from which it gets its name. Like other tits, in spring it sings a different song with various changes in rhythm. It nests in cavities and man-made nestboxes in which it lays six to eight eggs speckled reddish-brown. A resident bird, it stays the winter, at most roaming the countryside in the vicinity of its nest together with other tits.

Parus atricapillus

The Penduline Tit (428), a small bird closely related to the other tits, is noted for its remarkable nest, a bag-shaped structure of plant wool with a tubular side entrance suspended from the thin outer branch of a willow, birch or aspen above water. It breeds irregularly twice a year, each clutch containing six to eight white, unspotted eggs. It is found only sporadically in Europe alongside water with reeds and in central Asia to Lake Baikal.

Remiz pendulinus

429

The Long-tailed Tit (429), closely related to the other tits, is a familiar small bird and very social. So social are these birds that they roost pressed tightly to one another and during the nesting season help feed the offspring of other birds. The clutch consists of seven to fourteen yellowish, inconspicuously spotted eggs, laid in a closed spherical nest, which the birds build twice a year. They breed in Europe and in the central belt of Asia as far as Japan. The Long-tailed Tit is resident and therefore occurs in a number of geographical races.

Aègithalos caudatus

430

The Nuthatch (430) is a very active bird, climbing expertly on tree trunks, head up as well as head down, with a loud whistling call which one would not expect of such a small bird. It has the odd habit of reducing the size of the entrance to the tree hole, in which it nests, with mud, and lining it with flat bits of bark. It is one of the new resident birds that stay in their nesting territory throughout the year. The Nuthatch is a very noisy bird. Its warning call is often heard especially during nesting. It is distributed from Morocco across all of Eurasia and forms twenty-five separate races.
Sitta europaea

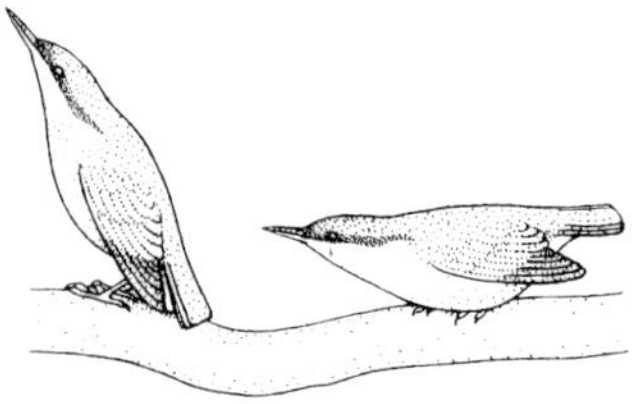

Part of the Nuthatch's courtship display when the male provokes the female with his back turned

431

432

The Velvet-fronted Nuthatch (431) is the most brightly coloured of the nuthatches, with blue
Sitta frontalis
head and back, black forehead and red beak. It inhabits the deciduous woods of southeast Asia.

The Short-toed Treecreeper (432) is one of five species of songbirds that climb only on tree
Certhia brachydactyla
trunks. They are small, brown-coloured birds related to tits. When they climb they always start at the bottom of the trunk and work their way upward, propping themselves with their tail feathers, which, like the woodpeckers', are pointed and have a strong rachis. The Short-toed Treecreeper generally locates its nest behind dead bark and lays six or seven tit-like eggs twice a year. It breeds from western Eurasia to central Europe and the Balkans and in North Africa, in the Atlas Mountain region; from central Europe eastward it is replaced by the Treecreeper *(Certhia familiaris)*.

The Hill Mynah (433), like all other members of the starling family (Sturnidae), is a very social
Gracula religiosa
bird. It is noted for its great learning ability and for that reason is

433

435

436

a popular cage-bird in its native land (in India and Sri Lanka). It can mimic various sounds as well as human speech often better than parrots. It is found in mountain forests where it feeds mostly on fruits, sometimes also on insects.

The Starling (434) is a popular bird and because of its aggressive character and also aided by man it has spread practically throughout the whole world. It is a typical representative of the family, which comprises 111 species of robust birds, some as big as a Magpie. Most have a dark colouring with a metallic sheen. The Starling is a voracious feeder which on the one hand consumes countless insects, worms and snails but on the other is a pest of fruit orchards and vineyards. It forms pairs but is fond of nesting in larger numbers in holes in trees and buildings. Its great expansion was aided by man by the hanging of nest boxes, a popular habit in many countries since medieval days. The clutch consists of four to six pale blue eggs; at lower elevations the Starling regularly has two broods a year, with both parents tending the offspring. When they have fledged, flocks of young birds roam the countryside. Starlings congregate in large roosts in reeds for the night. In recent years they have been roosting on tall buildings in large cities. Northerly populations are migrant, but in European cities Starlings are an increasingly common sight even in winter.

Sturnus vulgaris

European migration routes of the Starling

The Silky Starling (435) is an Asian species with ash-grey upper parts, white breast and belly and black metallic wing quills. It breeds only in central and southern China. During migration it is occasionally seen in the Philippines.

Sturnus sericeus

The Crested Mynah (436) is one of six species of southern Asian birds larger than the Starling, coloured black and with a characteristic crest on the forehead. The Crested Mynah is common from central China to Malaysia. It was also introduced into North America, where there is now a sizeable population in the vicinity of Vancouver.

Acridotheres cristatellus

437

The Bali Mynah (437) is a rare and endangered species found only on the island of Bali. It was not discovered until 1910. In several of the zoos where it is kept attempts are being made to breed the bird. It feeds mostly on fruit.
Leucopsar rothschildi

The Superb Starling (438) is one of the most beautifully coloured of the forty-five African species of starlings. They have glossy black, blue or green plumage and glowing yellow eyes. This sociable bird is found in bushy savannas with acacias, in the tops of which it builds its ragged, domed nest. It obtains the animal life it feeds on mostly on the ground. It often occurs in the vicinity of human settlements. It is found in eastern Africa from Ethiopia and Somaliland to southern Tanganyika.
Spreo superbus

The Blue-eared Glossy Starling (439) is widely distributed from Senegal and Cameroon to Kenya and Ethiopia. It inhabits wooded areas and bush country and is the only African starling found in desert oases. It builds its nest of dry grasses in the holes of trees and dead, rotting trunks, and lays only two or three pale blue eggs thinly covered with brownish spots. It feeds on insects, mostly locusts, as well as various fruits, and also visits grain fields. It is a noisy bird and the large congregations roosting in thickets above water deliver deafening concerts.
Lamprotornis chalybaeus

438
439

440

The Emerald Starling (440) is the most beautifully coloured of the West African species. It is
Coccycolius iris
smaller than the European Starling and occurs in small flocks in the bush and in open savannas; it avoids woods. Nothing is known yet about its nesting habits. It feeds mostly on ants and has often been observed collecting dead insects in burned areas destroyed by steppe fires.

The Red-legged Honeycreeper (441) belongs to the tropical American family of honeycreepers
Cyanerpes cyaneus
(Coerebidae) numbering thirty-seven species of small birds with brightly coloured, glossy plumage, resembling sunbirds. The Red-legged Honeycreeper inhabits the tropical forests from Mexico to southern Brazil. It pierces the bases of flowers for nectar but also eats sweet fruit and insects which it catches in flight. It builds an open cup-shaped nest and lays two white, finely spotted eggs. The lovely black and violet-blue plumage with turquoise cap is worn by the male only during the courting period after which he dons an olive-green garb with yellowish underparts like the female's.

The Black-faced Honeycreeper (442) inhabits the tropical forests of northern South America
Dacnis lineata
as far as Bolivia and central Brazil. Of the group of honeycreepers and flower-piercers it is the one closest to the tanagers. It feeds mostly on sweet fruit, being fondest of bananas, and also catches insects. The nest is a thin-walled structure of fine fibres. The two blue eggs with brownish spots are incubated by the female, but the male assists her in feeding the young.

441

442

443

444

The Grey-backed White-eye (443), like the other eighty or so species of this family, has a ring
Zosterops lateralis
of white around the eye. The White-eye is found in Africa, southern and eastern Asia and in the whole of the Australian region. It feeds chiefly on small insects, but many species pierce flowers and sweet fruit and sip the sweet juices. It builds an open nest and lays two to four eggs, which have the shortest incubation period of all birds — a mere ten to eleven days.

The Bronzy Sunbird (444) is one of 108 of the most beautiful and brightly coloured birds of the
Nectarinia kilimensis
tropical and subtropical regions of the world, except for America. Like the hummingbirds they circle about flowers and sip nectar with their long, thin, downcurved tubular bill. Unlike hummingbirds, however, only the males have glossy and brilliantly coloured plumage. The Bronzy Sunbird is common in the African highlands of the equatorial belt. It forms pairs, like all sunbirds, and builds a closed nest with side entrance in the thin outer branches of trees.

The Myrtle Warbler (445) belongs to a large family of more than 120 small and active birds
Dendroica coronata
found only on the American continent from the Arctic Circle in Canada south to Argentina. Many are among the commonest birds in America.

445

446

447

They are popular with man for they help destroy insects, which are their chief food. The various species are found in all sorts of habitats, being common also in city parks and gardens. Birds from more northerly areas fly south in large numbers in the autumn. The Myrtle Warbler is found in coniferous woods from the forest boundary in Canada to the northern United States.

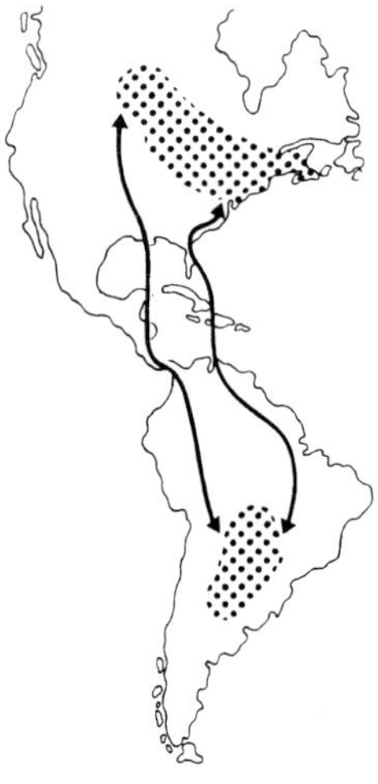

Migration route of the Bobolink

Typical birds of the American continent are the icterids (family Icteridae) numbering ninety-four species. Their colouring is reminiscent of the starlings and orioles but they are related to the tanagers and finches. They are as large as a lark or crow and most have a long, pointed, conical beak, the upper mandible being topped by a prominent swelling in many species. In most instances the sexes have similar colouring; the females are usually somewhat smaller. Many are excellent singers. Some feed mainly on nectar, fruit or seeds, others in great part on insects, and the largest species even eat small vertebrates. Icterids often nest in colonies, the majority building long, bag-shaped, hanging nests that are architectural marvels. Birds nesting in colonies are frequently polygamous, there often being more than five times as many females as males in such an aggregation. The group of cowbirds includes some that are social parasites, laying their eggs in the nests of other birds, but going about it differently from cuckoos. Unlike the latter, they are friendly with their hosts, often even outside the breeding season, and slip their egg into the host's nest at a convenient moment and without any disturbance.

448

The Baltimore Oriole (446), the orange and black oriole of North America, breeds as far north as southern Canada. Its return from its winter quarters in Central and South America is announced by a loud and clear musical whistle. The nest is a firmly woven basket-shaped structure with an entrance at the top, hung from the extreme end of a twig.
Icterus galbula

The Eastern Meadowlark (447), unlike most other icterids, is adapted to life on the ground, where it also nests. Its favourite haunts are damp places in prairies, meadows and fields in the eastern United States as far south as northern Brazil. It has brownish speckled upper parts and yellow breast with black V-shaped crescent. It feeds mostly on insects. The nest is located in tall grass.
Sturnella magna

The Bobolink (448) is a common and familiar bird of North America with finch-like body and beak. The male's nuptial plumage is white above and black below, the head is also black; this reversed colouring — light upper parts and dark underside — is unique amongst songbirds. The female is brownish with a striped head. The Bobolink nests in thick vegetation near the ground and usually has five to seven young. When they have fledged the birds form large flocks which visit grain fields, causing considerable damage by eating the grain. In autumn the birds migrate as far as Brazil and Argentina.
Dolichonyx oryzivorus

The Troupial (449) is widespread in Venezuela and outlying islands, where it is found in open woods and gardens. After the breeding season the males exchange their
Icterus icterus

449

450

beautiful orange and black garb for a dull olive green dress resembling the female's. The Troupial is often kept as a cage bird because of its ringing flute-like song.

The tanagers (Thraupidae) are a large family of birds resembling icterids and finches and numbering more than two hundred species distributed chiefly in the tropical zone of Central and South America. Some are brilliantly coloured birds. In most species the sexes have like colouring; only in some is the male more strikingly coloured. They feed on seeds, berries and varied fruits, some also on nectar and occasionally insects. The nest is cup-shaped and the clutch comprises two to four blue eggs covered with reddish-brown dots.

The Stripe-headed Tanager (450) is found in Cuba, where it is common in woods as well as city parks and gardens. It forms pairs and after the nesting season forms small flocks that visit orange plantations and nibble the fruit.

Spindalis zena

The Seven-coloured or **Orange-rumped Tanager** (451) is an east Brazilian species often kept as a cage bird for its beautiful colouring, which is the same in both sexes and can be admired throughout the year except during the moult. Loveliest are glittering green-blue head and glowing yellow patches on the tail coverts. The song, as in all tanagers, is moderately good.

Tangara fastuosa

Distribution of tanagers

451

452

The Scarlet Tanager (452) is distinguished by a marked difference in the plumage of the male
Piranga olivacea
and female during the breeding season. Whereas the latter is olive-green with yellowish underparts, her mate has a brilliant coat of scarlet with black wings and tail. It breeds in the deciduous and mixed woods of eastern North America north to southeastern Canada and winters in tropical South America.

Next to the flycatchers and warblers the family of finches (Fringillidae), numbering about 450 species, is the largest in the bird kingdom. They have a short, conical, hard beak for most are seed-eaters. The nest is located in thickets, trees or even on the ground. The birds form pairs, but outside the breeding season are very gregarious. They are distributed throughout the world except for Madagascar and Australia.

The American Goldfinch (453) is common throughout North America. The male is canary
Spinus tristis
yellow with black forehead and cap and black wings. It is found in open country with trees and thickets, in orchards and in tree avenues.

The Redpoll (454) occurs in several races in the tundras of Europe, Asia and North America.
Acanthis flammea
Farther south it breeds in England, in the Alps and in the mountains marking the border of Bohemia, where it has been spreading to low

453

454

456

elevations in recent years. Some years northern populations invade the whole of central Europe in vast numbers in autumn and winter.

The Goldfinch (455) is a familiar bird with bright colouring and pleasant song, which has made
Carduelis carduelis
it popular as a cage bird. It is found throughout Europe except in the wooded and northern areas, in Asia as far as Lake Baikal and in all of North Africa. It has also become common in many parts of America, Australia and New Zealand, where it was introduced by man.

The Crossbill (456) is a bird without a home, for it is continually on the move from one coniferous
Loxia curvirostra
forest to another, depending on the local crop of cones, for the seeds are the mainstay of its diet and are also fed to its offspring. The flight paths often cover distances of several thousand kilometres. The Crossbill is the only songbird that breeds regularly in winter, even at temperatures

457

45

45

of minus 35°C. It is found in the Atlas range in Africa, in the whole of Eurasia and in North America.

The Red-mantled Rosefinch (457) is a secretive bird of the high mountains of Tien Shan in central Asia. It nests in the highest pine-wood belt and barren stony parts. The male's head, breast and back are coloured pink tinged with silver.
Carpodacus rhodochlamys

The Bullfinch (458) attracts one's attention mainly in winter when the bare bushes are dotted with bright spots of red (the males' bellies), as small groups of these birds feed on seeds and berries. It commonly nests in coniferous and mixed woods, both in mountains and at lower elevations, as well as in larger city parks and gardens. It is distributed practically throughout all of Europe and in the forest belt in Asia all the way to Japan. The nest is generally hidden in young spruce thickets. It has two broods a year and the four or five eggs in each clutch are coloured azure blue with rufous dots and ash-grey markings. Birds from northern areas fly farther south in winter; they are distinguished from their local counterparts by their larger size and friendly, trusting nature.
Pyrrhula pyrrhula

The Long-tailed Rosefinch (459), the only species in this genus, has a long, graduated tail edged with a broad white border and rose-coloured breast. Its range extends from southern Siberia to Sakhalin, Japan and southern China, where it nests in thickets alongside rivers in mountains and foothills.
Uragus sibiricus

The Serin (460) was originally a Mediterranean bird which suddenly began to spread northward in the nineteenth century. Today it breeds regularly along the Baltic and has also been recorded breeding for the first time in England and southern Sweden. It is found in warmer localities (even in the mountains), but always near human habitations. The song is a continuous sibilant twitter
Serinus serinus

460

Spread of the Serin's range

461

generally delivered from a perch on a telephone wire or in flight, with the bird moving its outspread wings slowly like a butterfly. The Serin is fond of nesting in fruit trees. Birds from northern areas return to the Mediterranean for the winter but small flocks are staying the winter with increasing frequency in cities on weed-grown refuse dumps.

The Masked Hawfinch (461) a bird of the Far East, inhabits the deciduous woods of Manchuria, eastern China and Korea.
Eophona personata

The Hawfinch (462) is distinguished by its extremely large, robust bill which can even crack cherry stones. It nests in mixed and deciduous woods throughout temperate and southern Europe and locally in Asia as far as the Amur River.
Coccothraustes coccothraustes

462

463

The Slate-coloured Junco (463) is coloured a dark slate-grey with white belly and white outer tail feathers. It breeds in coniferous and mixed woods from the forest belt in Canada to the northern United States, in mountains as far as northern Georgia. The nest is well concealed and placed on or near the ground. In it are deposited four to five pale blue eggs sprinkled with reddish-brown.
Junco hyemalis

The Indigo Bunting (464) inhabits treeless or bushy country, usually nesting on the ground. The food consists mainly of seeds. The male Indigo Bunting, widespread from southeastern Canada to the middle United States, is a rich indigo blue; the female is olive-brown with faint streaks on the underparts.
Passerina cyanea

464

The Woodpecker Finch *(Cactospiza pallida)*, **one of the group of Darwin's finches, relatives of the buntings, skilfully uses a long thorn to extract insect larvae from crevices**

466

The Reed Bunting (465) is found beside water and swampy sites with reed beds and thickets throughout most of Eurasia. During the breeding season the male delivers his monotonous grating song from an elevated perch at noon on a hot summer's day when other birds are silent. The nest is well concealed in grass tussocks.
Emberiza schoeniclus

The Lapland Bunting (466) has a circumpolar distribution in the arctic and subarctic tundra. When singing it often flies up in the air like a pipit.
Calcarius lapponicus

The Cardinal (467), a familiar North American bird, makes its home in the southern and eastern United States southward to Honduras. It is found in parkland, gardens and groves. In the southern parts of its range it has as many as three broods a year. The birds pair for life.
Richmondena cardinalis

The Red-crested Cardinal (468) is common in the South American pampas with woods and thickets. Its range extends from southern Brazil and Bolivia to northern
Paroaria coronata

467

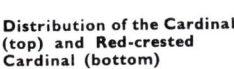

Distribution of the Cardinal (top) and Red-crested Cardinal (bottom)

468

469

470

Argentina. Unlike northern species of cardinals the male and female have like colouring.

Showing practically no differences in appearance from the finches is the last family of songbirds — the weavers (Ploceidae), containing more than 250 species. They, too, have short, robust bills — for they feed primarily on seeds — and brownish plumage with black, white or yellow markings. In some of the tropical groups the males have a bright garb with red, yellow and green areas. The weavers have social habits that are often very conspicuous, for example, during the breeding season the birds gather at dusk in large groups that make a deafening noise with their chirping and chattering chorus. House sparrows congregating in the trees of illuminated parks and city streets often keep this up until dawn, only then dispersing and returning silently to their respective nests. Weavers generally nest in colonies, either in huge structures measuring several metres and containing several hundred nesting chambers, or in individual spherical nests placed several on one tree or scattered in thickets and tall grass over a large area. True weavers are often master architects, weaving superb structures of long fibres with short or long tubular entrances. The type of nest and its location is an unfailing means of identification of the species. Even sparrows, which build their untidy nests of dry grass and feathers in widely varied cavities and crevices, have in places retained the habit of building spherical nests together in trees.

Most weavers are resident birds and remain in their nests throughout the year. Only in the autumn, when grain and grass seeds are ripe, do they make group forays into the fields. African weavers form huge aggregations of as many as several million birds. Weavers are characteristic African birds, though they are also found in Australia, Europe and Asia.

The House Sparrow
Passer domesticus

The House Sparrow (469, 470), found in Europe, Asia and North Africa, has become cosmopolitan, thanks to man, with whom it is closely associated — a troublesome companion wherever it occurs. It forms pairs and where possible nests together with at least several other pairs. It uses a separate nest for sleeping and as a place of concealment. During the courting season several males can often be seen hopping with drooping wings and loud chirps around a female, which eventually tries to fly off. The Sparrow has two or three broods a year and lays four to six eggs densely covered

Male Sparrow performing his courtship display

471

with grey-brown spots. Both parents tend the young, feeding them only insects the first few days.

The Tree Sparrow (471) is found in Eurasia, except for southeastern Europe and southern Asia, as far as India. Both sexes have a chocolate-brown cap and are indistinguishable from each other. The Tree Sparrow is found far from human settlements along tree-lined highways and in fields, but also occurs in city outskirts and parks. It is fond of nesting together with others of its kind, placing its nest in tree holes and nest boxes as well as in holes in the walls of sand and clay pits.

Passer montanus

The Rock Sparrow (472) is a thermophilous sparrow found in the Mediterranean, including northwest Africa, in the Middle East and in central Asia. It nests in small colonies on dry, barren slopes in mountains and in steppes, placing its nest in holes in mud and rocky banks.

Petronia petronia

472

473

The Taveta Golden Weaver (473), a small weaver the size of a sparrow, is found in savannas of Kenya and Tanzania. It nests in colonies being partial to thorny thickets, near coast to coconut palms. Its bag-shaped communal nest with entrance on the underside may contain several nesting chambers and is attached to end twigs.

Textor castaneiceps

The Red Bishop (474) occurs in several races throughout Africa south of the Sahara. It is the prettiest weaver of all, found in small colonies in grassy savannas. It begins nesting after the rainy season, at which time the males exchange the pale brown garb resembling the females' for a beautiful black and red dress. It nests in large numbers in tall grass, suspending the loosely woven nest about one metre above the ground. The clutch consists of three to six blue-green eggs, which may often be glimpsed through the thin walls of the nest. When the young have fledged the birds roam the countryside in flocks together with many other species.

Euplectes orix

474

475

The Black-headed or **Village Weaver** (475) is one of the most numerous species of weaver. It is found south of the Sahara in widely varied habitats near water, in savannas, in forest clearings, as well as near human settlements, in short, wherever the large trees can support its huge colonies. The nests, suspended close beside each other on the twigs and branches, are generally woven of palm fibres, obtained by the bird's nipping the frond and tearing off a long strip by flying off with it in its beak.

Ploceus cucullatus

Trees laden with weavers' nests (476) are a typical part of the African landscape. The nests may be suspended separately or in large clusters, or else the nest may be a tremendous communal structure with hundreds of holes opening on to the nesting chambers. Such colonial nesting has led to polygamous relations in some species and to social parasitism in others.

The Yellow-mantled Whydah or **Widow-bird** (477) is one of a group of weavers where, in the breeding season, the males are garbed in black and sport long, broad tail feathers that stream behind them like ribbons. They are polygamous birds with four to twelve females to one male. The Yellow-mantled Whydah inhabits the grasslands of tropical Africa. It nests in grass, building a sturdy, closed structure with side entrance.

Coliuspasser macrourus

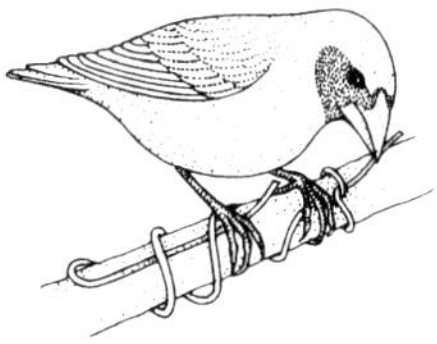

Initial stage in the building of a weaver's nest

Mouth of a young Gouldian Finch with phosphorescent nodules and coloured spots

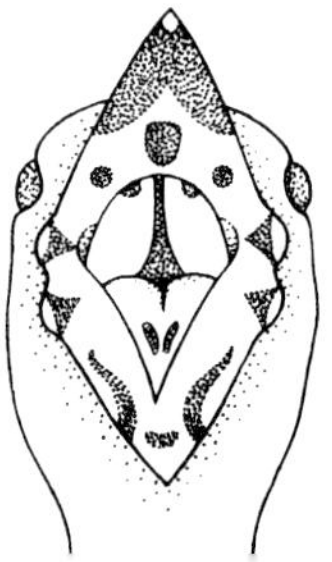

476
477

478

479

The Gouldian Finch (478) is one of 125 species of magnificently coloured weaver-finches, a true
Chloebia gouldiae
jewel and very popular with bird fanciers, inhabiting tropical Australia and Africa, where it is generally found near water in grassy savannas dotted with eucalyptus trees. The nest, a spherical structure, is placed in holes in eucalypti as well as in the open in thickets and in trees. The young have coloured spots and phosphorescent nodules inside the mouth which aid the parents in placing the food in the darkness of the nest.

The Java Sparrow (479) is a common, sociable bird of Sumatra and Java. It is often kept as
Padda oryzivora
a cage bird and has also been introduced into several places in Asia and Africa. It nests in the open in thickets and in trees as well as in various cavities, often even in human settlements. The nest is an untidy structure resembling the Sparrow's.

The Queen or **Shaft-tailed Whydah** (480) belongs to a group of nine African weavers all of
Tetraenura regia
which are social parasites. Each species places its eggs in the nest of only one species of waxbill to which it is so closely tied that the adult whydah has an identical call, the eggs, instead of being spotted, are pure white like those of the hosts, and the young are alike, demand food in the same manner, have the same growth period and even have the same bright mouth markings and nodules. The Queen Whydah is at home in South Africa and parasitizes the Violet-eared Waxbill *(Granatina granatina).*

The Zebra Finch (481) is one of the commonest of cage birds. It multiplies extremely well in
Taeniopygia guttata
captivity and has been bred in a wide range of colour mutations. It is a very common, social bird in grassy areas with bushes and trees in the

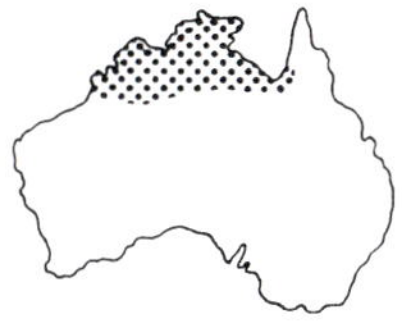

Distribution of the Gouldian Finch

Distribution of the Queen Whydah

480

81

482

interior of Australia. The closed, pear-shaped nest (several on a single bush) is used as sleeping quarters outside the breeding season.

The Long-tailed or **Shaft-tailed Weaverfinch** (482) is an elegant and very social Australian weaver, found in grassy savannas with eucalyptus trees, in the tops of which it generally nests. In dry areas it often occurs together with other birds in the vicinity of farms, where it has a dependable water supply. Birds pair for life and nest in scattered colonies. The nest is a spherical structure with an entrance tube about twenty centimetres long. This weaverfinch begins breeding at the start of the rainy season and generally has two broods; in dry years it does not nest at all. It feeds on seeds and insects, preferring ants and termites. When it encounters another bird of the same species it has the odd habit of nodding its head up and down. It, too, is a popular cage bird.

Poephila acuticauda

BIBLIOGRAPHY

Alexander, W. B.: Birds of the Ocean. New York 1954
Ali, Sálim; Ripley, S. D.: Handbook of the Birds of India and Pakistan. Bombay, London, New York 1968–1973
Austin, O. L.: Birds of the World. London 1963
Bannerman, D. A.: The Birds of West and Equatorial Africa. Edinburgh 1953
Bauer, K.; Glutz, U. von Blotzheim: Handbuch der Vögel Mitteleuropas. Frankfurt 1966
Benson, C. W.; Brooke, R. K.; Dowsett, R. J.; Irwin, M. P. S.: The Birds of Zambia. London 1971
Berndt, R.; Meise, W.: Naturgeschichte der Vögel. Stuttgart 1959–1966
Bruun, B.; Singer, A.: The Hamlyn Guide to Birds of Britain and Europe. London 1970
Cave, F. O.; MacDonald, J. D.: Birds of the Sudan. Edinburgh–London 1955
Chapin, J. P.: The Birds of the Belgian Congo. New York 1932–1953
Delacour, J.: The Waterfowl of the World. London 1954–1956
Dementiev, G. P.; Gladkov, N. A.: Birds of the Soviet Union. Moscow 1951–1954
Die Neue Brehm-Bücherei (Vögel). Wittenberg–Lutherstadt
Falla, R. A.; Sibson, R. B.; Turbott, E. G.: A Field Guide to the Birds of New Zealand. London 1970
Gilliard, E. T.; Steinbacher, G.: Knaurs Tierreich in Farben. Vögel. Munich–Zürich 1959
Grzimeks Tierleben, Vögel. Zürich 1968–1970
Hanzák, J.: The Pictorial Encyclopedia of Birds. Prague 1974
Hudec, K.; Černý, W.: Fauna ČSSR – Birds. Prague 1972
Iredale, T.: Birds of New Guinea. Melbourne 1956
Johnson, A. W.: The Birds of Chile. Buenos Aires 1965–1967
Makatsch, W.: Wir bestimmen die Vögel Europas. Radebeul 1966
Mauersberger, G.: Urania Tierreich – Vögel. Leipzig–Jena–Berlin 1969
Meyer, R. de Schauensee: The Birds of Colombia. Narberth 1964
Peterson, R. T.: A Field Guide to the Birds. Cambridge, Mass.' 1947
Peterson, R. T.; Mountfort, G.; Hollom, P. A. D.: Die Vögel Europas. Hamburg–Berlin 1954
Slater, P.: Field Guide to Australian Birds. Edinburgh 1971
Smythies, B. E.: The Birds of Borneo. Edinburgh–London 1960
Stegman, B. K.; Ivanov, A. I.: Birds of the Soviet Union – Key. Moscow–Leningrad 1964
Vaurie, Ch.: The Birds of the Palearctic Fauna. London 1959–1965
Voous, K. H.: Atlas of European Birds. 1960
Whistler, H.: Popular Handbook of Indian Birds. Edinburgh 1963
Williams, J. G.: The Birds of East and Central Africa. London 1965
Witherby, H. F.; Jourdain, F. C. R.; Ticehurst, N. F.; Tucker, B. W.: The Handbook of British Birds. London 1952

INDEX OF COMMON NAMES

(Bold figures refer to numbers of illustrations)

INDEX OF LATIN NAMES

(Bold figures refer to numbers of illustrations)